WITHDRAWN
UTSA LIBRARIES

MAKING ONLINE NEWS

Digital Formations

Steve Jones
General Editor

Vol. 49

PETER LANG
New York • Washington, D.C./Baltimore • Bern
Frankfurt am Main • Berlin • Brussels • Vienna • Oxford

MAKING ONLINE NEWS

The Ethnography of New Media Production

Edited by Chris Paterson & David Domingo

PETER LANG
New York • Washington, D.C./Baltimore • Bern
Frankfurt am Main • Berlin • Brussels • Vienna • Oxford

Library of Congress Cataloging-in-Publication Data

Making online news: the ethnography of new media production /
edited by Chris Paterson, David Domingo.
p. cm.—(Digital formations; vol. 49)
Includes bibliographical references and index.
1. Online journalism. I. Paterson, Chris. II. Domingo, David.
PN4784.O62M35   302.231—dc22   2008010857
ISBN 978-1-4331-0214-1 (hardcover)
ISBN 978-1-4331-0213-4 (paperback)
ISSN 1526-3169

Bibliographic information published by **Die Deutsche Bibliothek**.
Die Deutsche Bibliothek lists this publication in the "Deutsche
Nationalbibliografie"; detailed bibliographic data is available
on the Internet at http://dnb.ddb.de/.

The cover photograph of the Associated Press New York newsroom
is by Richard Drew, and was provided courtesy of the Associated Press.
Cover design by Clear Point Designs

The paper in this book meets the guidelines for permanence and durability
of the Committee on Production Guidelines for Book Longevity
of the Council of Library Resources.

© 2008 Chris Paterson & David Domingo
Peter Lang Publishing, Inc., New York
29 Broadway, 18th floor, New York, NY 10006
www.peterlang.com

Printed in the United States of America

Table of Contents

Acknowledgments vii

Foreword... ix
    Nora Paul

Introduction: Why Ethnography? 1
    Chris Paterson

PART ONE: Researching the Changing Nature of Media Production

1.   Inventing Online Journalism: A Constructivist Approach to the
    Development of Online News... 15
    David Domingo

2.   Ethnographic Media Production Research in a Digital Environment 29
    Roel Puijk

PART TWO: New Media, New Routines?

3.   News Production in an Irish Online Newsroom: Practice, Process,
    and Culture... 45
    Anthony Cawley

4.   Print and Online Newsrooms in Argentinean Media: Autonomy and
    Professional Identity 61
    Edgardo Pablo García

5.   News Tuning and Content Management: An Observation Study of Old
    and New Routines in German Online Newsrooms 77
    Thorsten Quandt

6. Maximize the Medium: Assessing Obstacles to Performing Multimedia
    Journalism in Three U.S. Newsrooms 99
        Jody Brannon

7. When Immediacy Rules: Online Journalism Models in Four Catalan
    Online Newsrooms 113
        David Domingo

8. Online Journalism in China: Constrained by Politics, Spirited by
    Public Nationalism 127
        Johan Lagerkvist

9. Do Online Journalists Belong in the Newsroom? A Belgian Case of
    Convergence 143
        Vinciane Colson & François Heinderyckx

PART THREE: Reinventing Journalism?

10. Ethnography of Newsroom Convergence 157
        Jane B. Singer

11. The Active Audience: Transforming Journalism from Gatekeeping to
    Gatewatching 171
        Axel Bruns

12. The Routines of Blogging 185
        Wilson Lowrey & John Latta

Epilogue: Toward a Sociology of Online News 199
        Mark Deuze

References 211

Contributors 229

Index 233

Acknowledgments

This project was conceived of several years ago, in part as a publication originating from the IAMCR[1] Working Group for Media Production Analysis where some of the work here was first presented. Authors have come and gone, David Domingo joined the project as coeditor, and finally with the blessing of Steve Jones, editor of the Peter Lang Digital Formations series, the project has come to fruition. Special thanks are due to the researchers writing in the book who joined the project in those early stages and patiently stayed aboard.

We want to underline the effort and passion of each of the authors of the chapters in this collection of research. Their work in the field, observing online journalists in their newsrooms, listening to their worries and aspirations, understanding their routines and constraints, is what makes this book an original contribution to the study of the Internet as a news medium.

We'd like to thank the people at Peter Lang for believing in a book based on hard and long worked empirical research. We are especially grateful to editors Mary Savigar and Sophie Appel for their ability to make the production of the book a nicely streamlined process.

We are also thankful to Arlecia Simmons for her thorough work proofreading the first manuscript of the book and her suggestions as an interested reader, far beyond her editing duties. This acknowledgment extends to the School of Journalism and Mass Communication at the University of Iowa for supporting Arlecia's research assistant position.

David would also like to express his gratitude for the permanent support, and intellectual and spiritual inspiration of Silvia.

NOTES

1.   International Association for Media and Communication Research

Foreword

Nora Paul

Director, Institute for New Media Studies, University of Minnesota

In this long overdue collection of online journalism research, the focus is placed on the newsrooms where a new news medium is being developed. These studies of journalists, struggling with uncertainties, dealing with new market pressures, and creating new journalistic conventions provide much-needed insight into the changes the Internet has (and has not) wrought on the work and practice of journalism.

The Internet is certainly not the first "disruptive technology" to hit newsrooms. Other innovations have radically transformed the news media business before, creating wholly new markets or destroying existing markets for other technologies, as Joseph Bower and Clayton Christensen theorized in 1995. The change from hot to cold type in the late 1970s fundamentally changed the process by which the newsprint page was constructed. The shift to cold type production led the way to the digitized newsroom, the move from clip files to data-based news archives, and ended the era of reporters banging out copy on their trusty Royal typewriters. The assembly line status quo of the newsroom was disrupted, responsibilities were reassigned and merged, and issues of quality versus efficiency were raised as tasks were bundled and assigned to one person.

No, the Internet is not the first disruptive technology that news organizations have contended with. But in terms of disruption, it may be the most widereaching.

Where the move from hot to cold type fundamentally changed the "back shop," and pagination changed the final stages of story production, the adoption of Internet technology as a new production, packaging, and distribution channel for news and information has fundamentally challenged every aspect of media organizations. New newsroom organizational structures and positions are being created to marshal the flow of copy to multiple distribution channels. New business models are being explored as Internet-based companies are competing with traditional media in the advertising market. New deadlines are being set. Even the very definition of what is "news" and who is a "journalist" is being reexamined.

Industry commentary and academic research has explored many of the areas where the Internet has impacted news organizations. There are examinations of the need for and viability of new business models. There is discussion about and content analyses of the ways online news is leveraging the opportunities for new ways to craft information. Critical essays about ethical issues in the online environment, surveys of professional attitudes and the new news consumer's behavior and preferences, opinions about how and why all sorts of online tools must be used in news organizations—these are all readily available.

What has been missing is careful examination of those at the center of the Internet's impact on the newsroom—the daily work of journalists, their work routines, and their values. There is certainly no shortage of anecdotal evidence of the Internet's impact on journalists' routines—just join any professional discussion list and a wide array of the issues will be readily raised.

But what has been missing is a dispassionate look at the work of journalists forging new ways to practice their craft in light of the technological, economic, and philosophical changes and pressures within newsrooms. How do journalists deal with these changes? How have they had to adjust their notions of their craft in light of new expectations (continuous deadlines, "beyond the newsroom" contributors, shifting priorities in news coverage)? Which traditions have they held on to, and what has needed rethinking?

Researchers in this book have entered the newsrooms, as sociologists in the 1970s did, to examine what is new, if anything, about online journalism. In this collection of studies, ethnographic research methods have been successfully applied to offer a fresh and thorough look at the actual evolution of Internet news.

A particular strength of the volume is its truly international scope. Argentina, Belgium, China, Germany, Ireland, Spain, and the United States are all represented in the studies of journalists working in online newsrooms. To date, too much of the research on the impact of the Internet on journalism is U.S.-centric. This collection lets us explore the unique, and common, challenges with which journalists worldwide are grappling.

The studies in this book provide an important baseline for journalism research. Future scholarship on the Internet will use these in-depth examinations of journalistic practices in the beginning years of online news production as a foundation to compare evolving attitudes, routines, and sense of professionalism.

The book also provides a fascinating insider's view of the work of journalists. Both news practitioners and news consumers will find revealing descriptions of the people who are committed to producing daily journalism in this time of change and uncertainty. Finally, online journalism instructors and students will, for the first time, have a tool to go beyond the theorizations of the opportunities of the Internet and understand the constraints real online newsrooms face in the process of exploring the potentials of online news.

It has been said that journalism is the "first take on history." In this time of historic change for the business and practice of journalism, this volume delivers valuable insights into the work of the reporters, photographers, editors, and producers on the frontlines.

INTRODUCTION

Why Ethnography?

Chris Paterson

We use, practice, and study online journalism. It influences us in untold ways. Every year more of us turn to the Internet for news and pay less attention to the media of old. We suggest in this book that for a phenomenon that has become so ubiquitous in the lives of all but those in the least connected parts of the world, we know virtually nothing of online journalism. Scholars examine what it has to say, what those producing it say about it, what its political and cultural influence is, and how audiences relate to it—but we still know little about what matters most: its construction. Is new media journalism really a new form of journalism? If so, how?

This collection seeks to take some fledgling steps toward understanding what is at the heart of news in new media: the process of online news production. Online editors often complain that they are exploring the Internet as a new territory without a map, and educators say that they have a hard time explaining to their students the work routines of online newsrooms. This book offers tools for both of them; it will help to make informed decisions about the nature of the field, and to describe in detail developments in online news production and work practices. Perhaps most crucially, we hope it will aid understanding of all the things online journalism is not—despite utopian predictions for it, stemming from the earliest days of the Internet.

For scholars, this book is an invitation to follow the path of ethnography and counter the technologically deterministic approaches behind many studies of online news. Research can no longer take for granted that the Internet will change journalism immediately and dramatically. In fact, media gurus still often do, as lately with the debate about citizen journalism, and ethnography is the best antidote: any technological development is embedded in an adoption process where social subjects make conscious or unconscious decisions that an observer can trace.

The research and theorization presented in this volume vary slightly in approach, but are united by an understanding that our "shared reality" (Berger &

Luckman, 1966; Tuchman, 1978) is increasingly shaped by the production practices of online media professionals, and that it is impossible to comprehend the nature of that manufactured reality without getting to the heart of the manufacturing process and the shared culture of the manufacturers. Literature reviews of online journalism research (Kopper, Kolthoff, & Czepek, 2000; Boczkowski, 2002; Domingo, 2005) suggest that studies have concentrated more on content, professional profiles and attitudes and audiences than on the production routines and context. Deuze, Neuberger and Paulussen (2004) stated that there is a clear distance between the ideals shared by online journalists and their actual practices, but observed that little empirical evidence had been published about the reasons for this distance.

Ethnography in (Online) News Production Research

It is our guiding premise that only ethnographic methodologies derived from anthropological and sociological traditions can come close to providing an adequate description of the culture and practice of media production, and the mindset of media producers. As Schlesinger (1980, p. 363) explained, the ethnographic method of news production research makes available "basic information about the working ideologies and practices of cultural producers," and provides the possibility of observation—informed by theory—of the social practices constituting cultural production. This is impossible with other methods, such as surveys or web content analysis—the dominant modes of online news research. Cottle (2007, p. 2) more recently argued that ethnographic studies of news production "help to reveal the constraints, contingencies and complexities 'at work' and, in so doing, provide the means for a more adequate theorization of the operations of the news media and the production of the discourses 'at play' within news media representations."

The shift away from sociological analysis of news production—common in the 1970s—was unfortunate and premature. The title of this book pays a tribute to that work, exemplified by Tuchman's *Making News* (1978). Without those early ethnographic investigations of news production, our understandings of journalism would be limited to what little we are able to glean from the observation of news content, or from what journalists say they do (which as any ethnographer soon discovers, often varies significantly from what they actually do). Among the important and influential large scale sociological studies of news production prior to the 1980s were Buckalew (1970), Warner (1970),

Epstein (1974), Altheide (1976), Schlesinger (1987), Tuchman (1978), Gans (1980), Golding and Elliott (1979), and Fishman (1980). But the relevance of these earlier works of news sociology is becoming marginal, for modern newsrooms—even the few still producing exclusively for "old media" channels—bear an ever decreasing resemblance to newsrooms of the late 1960s and early 1970s.

In their revisitation of the early research, Tuchman (2002) and Schudson (2003) commented on the benefit of vivid and thorough descriptions of work routines that explained journalism; but also noted the limitations of the ethnographic methodology. The newsroom-centric approach can be found lacking in a modern context increasingly dominated by the source-journalist relationship, which includes both the public relations professional's ability to manufacture news, and the dominance of news agencies in agenda setting (McManus, 1994; Manning, 2001; Paterson, 2007). Nonetheless, newsrooms are the actual space for decision making in the development of online journalism, where genres, routines, values, and products are tested and created.

This book presents the work of a "second wave" of ethnographers (Cottle, 2000) who are aware of the challenges of the new context and have a particular interest in technological innovation. We think this is the first time theorics of new journalism have been presented in conjunction with, and in the context of, a collection of many of the most substantial ethnographic research projects on Internet journalism conducted in recent years. The purpose of this book is not only to delineate how news production for new media is different from that of traditional media, but also to ask if it is different. There is occasionally reason to doubt if online media is terribly unlike old media; the places most U.S. online news consumers go for their news, for example, closely model the traditional American broadcast network structure in many respects (Paterson, 2007). As much of the research in this volume demonstrates, additional research is desperately needed to test many of the utopian predictions once—and sometimes still—made for new technologies (Domingo, 2005).

And there are larger questions thrown up by the digitization of journalism and the evolution of a new breed of journalist. Too often academic research into online journalism has what Halloran (1981) would have termed a "conventional" or "administrative" motivation: to find more effective and efficient ways to make the product. As Singer notes in her chapter, some early online newsroom research especially was oriented exclusively toward identifying "best practices" for industry. The problem, of course, is that crucial questions about the product, and its place in society, may be ignored. Media work (and research about that work) is not performed in a vacuum, independent of its cultural and political context. To date, for example, little of this research begins to examine the role of new media as a site of societal power struggle or to suggest the extent to which Internet journalism reproduces the cultural/ political/economic influence over society of concern to critical scholars, and documented, for example, by Glasgow (Eldridge, 1995), Gitlin (1980), or Schiller (1989). Perhaps that is asking too much of an emerging form of inquiry, but if the big questions do not shape our research from the outset, they are too often forgotten.

Online journalism research to date has mainly proven that the Internet features such as interactivity or hypertext that were meant to revolutionize the way news is produced and consumed were never extensively developed (Chung, 2007; Oblak, 2005; Kenney, Gorelik, & Mwangi, 2000; Schultz, 2000; Massey & Levy, 1999), even though professionals shared those utopian visions (Deuze et al., 2004). Some researchers, however, admitted that their methodological approaches failed to explain the factors shaping this paradox (Jankowski & Van Selm, 2000; Kenney et al., 2000) and reviewers have pointed out that most of the studies have limited themselves to denouncing the distance between the ideals and the reality without adding historical perspective or social context (Deuze, 1999; Carey, 2005) and with a lack of the empirical data and theoretical frameworks necessary to interpret the causes and consequences of these trends.

Boczkowski, whose comprehensive ethnographic project on U.S. newsrooms is a rare exception to the rule, wrote that these approaches tended to "build analysis upon a usually taken-for-granted technologically deterministic matrix" (2002, p. 279). If surveyed online journalists are convinced that they form a new breed of journalists (Quandt et al., 2006; Paulussen, 2004; Singer, 2003; Deuze & Dimoudi, 2002; Neuberger et al., 1998), we need to enter the newsrooms to see to what extent this is the case in their daily routines.

Online news production research is a paradigm that is both immature and controversial. It has many variations, sometimes with little in common apart from a shared claim to the term "ethnography" (and a number of studies which are, for all intents and purposes, ethnographic do not make mention of that term). Moreover, finding a clear and consistent definition for "ethnography" in the literature of communications studies is a challenge. One text circularly states of "the ethnography of communication" that "its methods are mostly ethnographic" (Lindlof, 1995, p. 46). But Lindlof more helpfully points out that the term sometimes means almost any form of qualitative research, accounting for why it is slippery. Despite this lack of agreement over exactly what the method does—or should—entail, self-described ethnographers like the editors of this collection have a habit of evangelizing about the value of ethnography and calling into question understandings of technology and society which have not been informed by it.[1]

Other methodologists insist the term should apply only when a researcher observes another culture over a long period, as in the anthropological tradition. Observational research is typically conducted in conjunction with extensive interviewing, although the reverse is less often true. In addition, some form of exhaustive document research or analysis of texts created by the culture in question may accompany both of those methods.

Domingo's (2003) analysis of benefits and weaknesses of ethnographic methodology is useful for those considering the approach (see table 1). These will be helpful to any researcher weighing the costs, in time and resources, of serious ethnography against the drawbacks of the method and advantages of alternative methodologies. Domingo's suggestions restate and update those made by Schlesinger (1980) a quarter century earlier, making obvious how despite change in media technologies, proven methodologies remain relevant, as do the central questions about their application.

Table 1. Benefits and weaknesses of ethnographic methodology for online journalism research

Benefits	Weaknesses
– Gathers a huge amount of very rich firsthand data.	– Observation is time consuming and many times actors feel disturbed by the presence of the researcher.
– The researcher directly witnesses actions, routines, and definitions of technology and social relations.	– It is not always easy to set down everything that you witness. Technical actions are the most difficult.
– The researcher can gain a confident status with the actors, obtaining insiders' points of view.	– Actors may ask you not to quote a confession they have made to you.
– The researcher can witness conflicts and processes of evolution.	– Results should not be generalized right away and you have the risk of taking an anecdote as a rule.
– Analysis of the gathered data allows a comprehensive description of the social use of a technology and offers insights to understand the factors involved in its social construction.	– The researcher has to be self-reflective, aware of his/her own prejudices in order to avoid them negatively influencing the study.

Here we strive to lend some coherence to the set of new and established research practices laying claim to the title of "online news ethnography." In doing so, we suggest ways in which these variations might complement each other, and ways in which "online news ethnography" might move forward into important areas of which we still have only the most superficial sociological understanding. This editor comes to this task as an ethnographer of old media, frustrated by the lack of reliable sociological research in new media—especially online news production. Editor Domingo came to ethnography feeling deceived by online journalism's ideals—in the 1990s the Internet was said to change journalism for the better—which were challenged by the real-life conditions of online newsrooms he knew as a media consultant. The recurrent hype of convergence and participatory journalism in the 2000s confirms that the Internet still is a powerful digital myth (Mosco, 2004) that drives the imagination of troubled media managers looking for a brighter future (Domingo, 2006, p. 64). That is why understanding the inner forces in the evolution of online media is more necessary than ever.

So, What Does Online News Production Look Like?

I won't spoil a good story (ethnography thrives on rich narrative) by trying to summarize subsequent chapters here, but I can observe that the research presented here (along with the further research referenced by these authors) describes a grimmer world of online news production than the hype of the last fifteen years predicted. Generalizing the findings of ethnographic studies is not advisable, but the similarities across chapters from very different geographic (from Argentina to China) and temporal (from 1998 to 2007) settings is striking and suggests established patterns in the industry.

Some predictions have—mostly—come true: deadlines have disappeared in the world of online news production, usually, however, with negative implications for the quality of news; technical barriers to online journalism have steadily eroded, but that seems to have had little impact on either the extent of convergence between old and new media production, nor on the socio-cultural divide between old and new media journalists—which remains profound. That divide often, it seems, still amounts to a second-class status for online media professionals, which is surprising given the maturity of the field.

Perhaps most importantly, there is very little indication that online journalism is inherently better journalism for all its interactive and in-depth potential, for the *shovelware* phenomena (repackaging content produced for other media) continues unabated, with news agency, print and, to a lesser degree, broadcast journalists setting the agenda for websites. Online journalism, at least as described by the ethnographers here, remains largely the "passive journalism" McManus (1994) found to have become the norm in U.S. television newsrooms in the early 1990s.

This anthology seeks to bring together some of the most current theories of new media production with the most current research into that production. It will be for the reader to consider how closely the work practices described by the ethnographers correspond to the theories. The first short section in the book deals with the theoretical and methodological framework shared by the authors: media production sociology, the social shaping of technology, and ethnography. These authors seek to frame the essential challenges in understanding online journalism and suggest key lessons from the research to date. The second section presents empirical research on online newsrooms around the world, focusing on different aspects of production routines, and carrying out their examinations at different historical periods.

These ethnographic chapters offer detailed descriptions of online journalists at work and discuss how and why they produce online news the way they do. The research is based on direct observation of the newsrooms and in-depth interviews with online journalists, although the research approaches vary considerably. A third section completes the book with examination of the contemporary trends of convergence and participatory journalism, and these authors suggest avenues for exploration of these developing areas. While online news production may differ significantly from newsroom to newsroom, it appears to reconceptualize journalism in consistent ways. Several contributions reflect in more depth on these characteristics, and in his concluding chapter, Deuze takes on the difficult—and possibly contentious—task of suggesting overall lessons about new media journalism that emerge from this diverse collection of research.

A Call to the Ethnographic Trenches

As the Internet and the research surrounding it evolved, researchers have maintained a fixation with content and the bulk of the literature is informed by analysis of content or reception—but little by analysis of production; and almost none utilizing the thorough ethnographic methodologies advocated here

(Domingo, 2005). A key reason or the lack of such research is access (Garcia, 2004). As noted by Schlesinger (1980) and others, the process of gaining the access to conduct ethnography of media production can be uncomfortably challenging, and that access, when granted, can be tenuous.

Media outlets were never enthusiastic about giving unfettered and long-term access to visiting researchers, but the doors have closed tighter with the consolidation of corporate media ownership. Image-conscious companies see little value in exposing their practices to the critical scrutiny of social scientists (unless assured of a fawning portrayal, as with much of the research into CNN). One may hypothesize—perhaps research will soon prove or disprove this—that the tendency toward *shovelware* among larger online media outlets was something media companies preferred not to expose to public scrutiny.

Puijk gives thoughtful consideration to the difficulties of access, noting the increasing security culture in larger media organizations makes it easy for executives to prevent or limit a researcher's access, even when journalists are eager to cooperate. Since little is written about unsuccessful attempts to gain access to media organizations, we can only speculate on the problem. The ever-increasing conglomerate control of media production can only exacerbate access challenges, with individual newsrooms within conglomerates given less and less autonomy. So much of our culture and information now emanates from a mere few corporations, yet ethnographic research about information production—and control—within the journalistic "brands" of those corporations is sparse.

There are many practical reasons for the lack of ethnographic research into new media production as well. It is cheaper, and easier, to analyze the Internet from one's own desktop than to locate oneself for weeks or months within a production setting. Long-term ethnographic research takes time and, more crucially, money. Media ethnographers need to be able to make repeated visits to their research site, house and feed themselves there for the long term (or pay others to be there), and often pay for the transcription of hours of interviews—all prior to the actual analysis of data. Few academics receive the funding or time from their institutions for such work, few funding agencies show enthusiasm for it, and few postgraduate degrees provide support for students to undertake such projects. While few in the academy will dispute the value of such research, institutional efforts to facilitate it are rare.

As Gans pointed out, reflecting on the lack of newsroom ethnography in recent decades, it is "very hard work, takes a long time, and can't be done while teaching fulltime. Thus, it is not functional for young people who need to produce a lot of articles to get tenure. Indeed, nothing can beat numbers-crunching on the computer for that" (Reese, 1993). Nearly all "field research" is valuable, but many studies described as ethnographic research have not included direct and intimate exposure to a subject culture lasting weeks, months, or years. We are pleased to have included here a diverse group of studies that can reasonably be considered (in various ways) thorough, substantial, and revealing ethnographies of online news production in its varied forms. But as with any extensive empirical research appearing in an anthology, you will read in these chapters just summaries and significant findings from large projects which, in most cases, have been or will be published in a more detailed form (the contact details of each author are provided at the end of the book, should you wish to learn more).

Readers will find that even within the limited scope of this volume, approaches to ethnographic research vary considerably. They range in approach from in-depth interviews to fairly unstructured but long-term observation to a highly systematic exercise in data gathering and analysis. The narratives recounting the findings of these studies similarly range from the deeply descriptive and literary to the highly quantitative. But all benefit from the key attribute of ethnographic work—direct and profound contact with the news workers and, in most instances, their working environment and culture. Within the range of studies in the book then, future practitioners of the ethnographic approach have the opportunity to assess different variants of the method and weigh their respective advantages and disadvantages.

We believe there is also a particular value in this multinational collection of research. As Quandt observes in his chapter, findings from one country do not necessarily apply in another. But we could also observe that findings in one country can, and probably should, be tested in other countries. This point is especially resonant in the case of the United States, home of a massive online news industry, but home to surprisingly little ethnographic research on that industry to date. The value in collecting findings from many nations in this volume is not that universal truths about online journalism will emerge, but that a wide range of ideas about what online journalism may and may not be are on offer, and it remains for future scholars to test the durability of these observations in the realms of online production they know best, and those which they have access to.

Opening each chapter, readers will find an "editors' note" that contextualizes the contributions of the author to the collection and serves as a quick browsing guide for the book. The reader will also notice a lack of references following each chapter. Rest assured this does not stem from neglect by the authors or editors, but from our wish to provide something we hope will be more helpful: a combined bibliography at the end of the book that might serve for all readers as a further reading list on new media journalism and the methodologies used to research it. This is not an attempt to provide a comprehensive and inclusive list of all publications on these topics, as this list only contains references used by the authors and editors of this book in their chapters. It should be a useful resource, though, for scholars beginning investigation of this field, or for practitioners wishing to discover further academic research about their field. We find it an especially novel bibliography, for it includes examples of research conducted and published outside the U.K.-U.S. nexus which literature reviews are too frequently confined to.

The assemblage of a multinational anthology carries the benefit of a greatly increased diversity of views on our common topic, but also the predicament of translating ideas from one language to another or conveying the nuances of ethnographic discovery in a foreign tongue (or a foreign style). Many of the studies described here were conceived, conducted, and originally presented in languages other than English, and so the editors have sought to work with the authors to produce chapters containing sufficiently clear English at a level that will serve students, academics, and media professionals alike, while also adequately preserving the author's own style of presentation. If lapses in clarity remain, the fault lies with the editors, not with the researchers whom we have coerced to work outside of their native language.

We hope this book will inspire all of us in the community of scholars studying and teaching about new media to get ourselves, our peers, and our students into the real world of new media production for extended stays, in order to generate the "thick description" (Geertz, 1973) still largely absent from our understandings of new media and online journalism. It is also our hope that it will inspire our colleagues in the business of new media production to welcome and facilitate in-depth examination of their emerging field, and use the data compiled here to examine their work practices and empower themselves with a better understanding of the challenges and constraints they face as they strive to develop an online journalism superior to the channels of news which precede it. Join the conversation at the book website: www.makingonlinenews.net.

NOTES

1. By way of example, Paterson has coordinated, since 1999, the Media Production Analysis Working Group of the International Association for Media and Communication Researchers, intended specifically to encourage research at the site of media production.

PART ONE

Researching the Changing Nature of Media Production

CHAPTER ONE

Inventing Online Journalism:
A Constructivist Approach to the
Development of Online News

David Domingo

This chapter presents an overview of theoretical traditions that can inform the research on online news production. From the sociological approaches of the 1970s to disciplines dealing with processes of technological innovation, Domingo argues that, in order to avoid technological determinism, researchers should have a constructivist perspective when they study changing environments such as online journalism. Ethnographic research of online news production is put in the context of journalism studies and a tool kit of concepts that can guide further research is offered.

EDITORS' NOTE

The research agenda in the field of online journalism has been dominated by studies produced in the United States, partly because of the leading and referential role of the country in the development of the Internet as a news medium. Studies in Europe and other regions of the world have usually followed theoretical and methodological proposals as they evolved in the American literature on the topic. Starting as early as the first media companies tested the Internet as a news medium, in the mid-1990s, we can identify three waves of research in online journalism (Domingo, 2005):

1. *Normative and prospective studies.* Researchers focused on building up ideal models for the development of online news based on the technical and communicative features of the Net (hypertext, multimedia, interactivity, immediacy/asincrony). Authors tended to overstate "the revolutionary character of online technologies and the Web" (Boczkowski, 2004a, p. 2), persuaded by the technological determinism that is inherent to capitalist societies since the industrial revolution (Katz, 2005): "Journalism is under-

going a fundamental transformation…. A set of economic, regulatory and cultural forces, driven by technological change, are converging to bring about a massive shift in the nature of journalism" (Pavlik, 2001, p. xi). These utopian proposals were very useful to delineate paths of innovation for the industry—as they still do in the 2000s with concepts such as convergence or participatory journalism—but they were unrealistic in describing the ideal models as necessary outcomes of online journalism.

2. *Empirical research based on the theoretical assumptions of the first wave.* Many studies of online news have concentrated on testing the ideal models, checking through website analyses or surveys the extent to which the models were being developed in online media. The early empirical results dramatically contradicted the normative perspective: Deuze summarized the basic results of this branch of research by stating that "most of the websites do not offer any 'online extra' in respect to the traditional version of the medium, they do not use hypertextuality, multimediality nor interactivity" (2001, p. 2). However, many researchers understood this result as an underdevelopment of online news and assumed that eventually the ideal model would be achieved. Furthermore, surveys of journalists detected "a gap between, on the one hand, online journalists' perceptions of the Internet's potential and, on the other hand, the actual use of interactive features" (Deuze, Neuberger, & Paulussen, 2004, p. 22). The striking phenomenon of professionals embracing the ideal model but not being able to put it to work could not be explained with the theoretical framework inspired by technological determinism or the methodology of the studies based in content analysis or quantitative surveys (Boczkowski, 2002).

3. *Empirical research based on a constructivist approach to technological change.* In an attempt to overcome the limitations of the previous perspective, researchers with a background in the sociology of technological innovation and communication history opted for qualitative methodologies to explore the reasons behind the current developments in online journalism (Singer, 1997; Brannon, 1999; Martin, 1998; Eriksen & Ihlström, 2000; Boczkowski, 2004b; Domingo, 2006). This wave is a shift from the others in several ways: first, the object of study changes from the effects of innovation to the process of innovation, with a constructivist perspective instead of deterministic; second, the ideal models are understood more as a factor that interacts with others than as a destination for online journalism; third,

researchers opt to analyze specific cases in depth to get closer to the decisions, routines and structures of online newsrooms in order to be able to describe the context and dynamics of the development of online journalism in each media organization. This exercise of "historicizing and localizing new media" (Boczkowski, 2004c, p. 146) and understanding innovation as an open process unlocks the assumption that the ideal models are a necessary goal for online journalism and helps the researcher explain the complex process through which professionals are defining a new news medium, highlighting the diversity of solutions—or understanding the reasons for homogeneity. It also helps researchers to have a critical perspective on the actual developments of online news and put the responsibility for the future of the profession back into the hands of journalists.

These three waves of research were born in the 1990s, but the weight of each one in terms of scientific production and consensus has shifted slowly from the first wave to the second. While in the late 1990s the trend in Internet research as a whole was driven toward the growth of social constructivist approaches that questioned the initial deterministic research (Lievrouw, 2004, p. 13; Wellman, 2004; Boczkowski & Lievrouw, 2007), the third wave in online journalism research is still far from becoming central.[1]

Some early literature on online journalism research (Deuze, 1998; Singer, 1998; Kawamoto, 1998) already urged an effort from scholars to explore old and new theoretical and methodological frameworks to better understand the ongoing evolution of this new form of news production. In the broader context of studies on the Internet as a social phenomenon, Jankowski and Van Selm warned that methodological innovation should not lead scholars to start research in the field from the scratch: "A frequent swan song is that Internet research is, in some way, 'different' from other forms of social science investigation and, therefore, requires unique, yet-to-be-redefined methods of study" (2005, p. 200). Following Williams, Rice and Rogers (1988, p. 13), they argued that new media research[2] should better build on conventional research methodologies and theoretical frameworks and explore alternative methods and designs from the ones assumed as standard. In fact, Jankowski and Van Selm (2005) point out that most of the innovations in Internet research are limited to the *micro* level of data gathering and analysis techniques, in many cases with the adaptation of existing methods to best manage the digital/virtual environment of the studies. Therefore, contributions in the *meso* (methodological development) and *macro* (theoretical development) levels are scarce: innovative research designs, consideration of different theoretical and disciplinary approaches, reflections on epistemological principles. This chapter intends to synthesize the contributions of the constructivist perspective to the meso and macro levels of online journalism studies.

The Sociology of News Production: Inspiration and Limitations

The sociology of news production has half a century of tradition in the study of newsroom routines and professional values, and has created a solid theoretical corpus to describe rules, roles, and processes and analyze their interrelations and consequences (see the introduction by Paterson). In this sense, it serves as an ideal guideline for the observation of the routines in online newsrooms. Moreover, empirical evidence on mass media news production can be compared to online routines in order to check for the variables of continuity and change. Nevertheless, in the last two decades scholars have pointed out that this research tradition has serious limitations for the analysis of innovation in the media. Most newsroom ethnographies lack historical perspective and overlook processes of change in news production routines (Schudson, 2000, p. 194). This never was the object of study of the sociology of news production, as the aim was identifying the organizational factors that shaped the news. Research concentrated on highlighting homogeneity and constants rather than on diversity and evolution (Cottle, 2000).

An effective strategy to overcome these limitations when analyzing online journalism as a developing phenomenon is to focus the researcher's attention on technology and ask how online newsrooms adopt technological innovations (Cottle, 2000, p. 33; Boczkowski, 2004b, p. 197). Heinonen (1999) suggested that online journalism research should acknowledge the *historicality* of journalism to be able to conceptualize change, something that even though the sociology of news production did not have in its research agenda was nevertheless implicit in the constructivist approach of the discipline. "Journalism is a social phenomenon: it emerged as a consequence of certain social (including technological and economic) developments and it is attached to certain cultural (including political) formations" (1999, p. 11).

Boczkowski (1999) and Cottle and Ashton (1999) defended the necessity for the sociology of news production to borrow theoretically solid conceptual frameworks to address the analysis of technological innovation in the newsrooms. From different standpoints, but with very similar approaches, science and technology studies and communication history offer fruitful tools that can "extend the valuable tradition of sociological studies of newsmaking" (Boczkowski, 2002, p. 278) to understand the emergence of the Internet as a news medium. In fact, the epistemological principles of all these three research traditions are rooted in the same origin: phenomenology and the sociology of knowledge (Berger & Luckmann, 1966) Each of the disciplines has obviously focused on different research questions until recently, but Boczkowski and Lievrouw (2007) detected an intensifying dialogue between studies on technological innovation and those centered on communication and the media. In the following section, useful concepts of the former are explored in order to contribute to strengthen the "intellectual bridges" (2007, p. 4) between traditions by enriching the sociology of news production.

Constructivist Research on Technological Innovation

A comprehensive analysis of the social adoption process of a technological innovation such as the Internet requires the understanding of technologies as a socially constructed multifaceted reality, and not a monolithic element that appears from nowhere and imposes its own logic to social actors such as media companies. Technological deterministic approaches assume that "the nature of technologies and the direction of change [are] unproblematic or predetermined" and they have "necessary and determinate 'impacts' upon work, upon economic life and upon society as a whole: technological change thus produces social and organizational change" (Williams & Edge, 1996, p. 868). Historical and empirical data have shown that "technological determinism is unsatisfactory because technologies do not, in practice, follow some predetermined course of development" (Mackay & Gillespie, 1992, p. 686). "There is never a single technical solution," argues historian Patrice Flichy (2006) offering some lessons of the past. Researchers need to study both "successes and failures" (2006, p. 188).

In the last decades of the 20th century, communication historians, innovation sociologists and anthropologists of technology have searched for an adequate research strategy to understand the processes that the invention of a new technology undergoes as well as the social adoption of such an innovation, as a way to overcome that simplistic approach. The common ground for these approaches is to understand technologies as a product of society: there is a social context where they are invented, which determines the "intention" of the researchers in developing them (Williams, 2003, p. 7), and a social context where they are adopted, in which users negotiate with the proposed definitions of the technology to adapt them to their needs and adapt themselves to the requirements of the technique usage. "Innovation is thus seen as a contradictory and uncertain process. It is not just a rational-technical 'problem-solving' process; it also involves 'economic and political' processes in building alliances of interests...with the necessary resources and technical expertise, around certain concepts or visions" for the use of a technology (Williams & Edge, 1996, p. 873).

In the case of online journalism, the Internet is an already existing technology adopted by media organizations for a new purpose that was not in the original plans of the 1970s researchers that created the network (Abbate, 1999): the diffusion of news and associated services. The context of the negotiation and definition of the Internet as a news medium are media organizations and, more specifically, online newsrooms, framed under the highly standardized set of rules and processes of the institution of mass media (McQuail, 2005).

Boczkowski summarizes the core of these disciplines in a "conceptual lens" that:

- historicizes new media;
- highlights the situated character of the practices that enact their construction and use;
- emphasizes the process dimension of these practices;
- pays special attention to restoring the visibility of material and social dynamics that tend to become less visible when new media become institutionalized; and
- accomplishes these goals through a methodological commitment to reaching intimate contact with, and detailed knowledge of, the phenomena under study. (2004c, p. 145)

The next sections provide the discussion on theoretical frameworks proposed by these diverse but converging group of research traditions.

Communication History

Historians have concentrated on identifying the big trends in the evolution of mass media in the last centuries and, within this framework, they have shown great interest for the analysis of communication technologies' invention and development. Their diachronic approach, focused on the "real agencies"—i.e., the actors and social factors involved—(Williams, 2003, p. 140), has let them ascertain that there are successful technologies and others that are a fiasco, as well as technologies that end up being used for other purposes than the ones initially intended. The history of technologies such as the radio or the Internet itself are the best antidote against technological determinism and the basis for the theoretical constructs of these scholars' regarding the relationship between technology and society.

Raymond Williams considers that a technology can be defined as a "social institution" that includes "technical inventions" (devices, artifacts), "techniques" (particular skills to use them), the "body of knowledge" to develop both devices and skills (in the scientific and engineering world) and, finally, the knowledge to use them in a particular social setting (1981, pp. 226–227).[3] Any technology is "the product of a particular social system." The scientific innovators usually take existing or foreseen social needs as the object of their research and look for technical solutions for particular social activities (Williams, 2003). In this initial stage, technology is already a social product: its features are usually based "on the representations that the designers [of the invention] share about the possible social uses and the most efficient strategies to commercialize the product. In other words, decisions are more social than technical" (Flichy, 1999, p. 34). As a consequence, the evolution of a technology is not predefined or foreseeable: it is "a process in which real determining factors—the distribution of power or of capital, social and physical inheritance, relations of scale and size between groups—set limits and exert pressures, but neither wholly control nor wholly predict the outcome of complex activity within or at these limits, and under or against these pressures" (Williams, 2003, p. 133).

Williams and Flichy's considerations shape a very suggestive broad theoretical framework, but it is difficult to transform them into systematic methods of analysis for the present situation, in which we have little historical perspective. Brian Winston (1996, 1998) proposed an operational model to analyze the real implications of the development of the Internet as a mass medium. For Winston, the invention and social adoption of the Internet is another step in a long, steady evolution.

Winston's efforts to systematize a model tend to simplify the social context of technological innovation, but he offers two useful concepts to make more visible the tensions in the process of consolidation and social adoption of a technology: *accelerators* and *brakes* of technological change (1996, pp. 21–25; 1998, pp. 3–15). Accelerators are "supervening social necessities" (1998, p. 6) that push forward technological innovation and social adoption of a new invention. Winston locates the accelerators in different social areas: they can be social groups' demands, innovations in related technologies or the industry need for new products, among others. A technical solution can be available, but it can only will be established as a technology when a social need promotes its use. Brakes are actions of specific social actors that try to slow down social adoption of a new technology. Winston defends the existence of a "law of radical potential suppression" (1998, p. 11); every technological innovation confronts social reactions to stop it, which gives existing institutions some time to adapt and change without losing their main attributes and power. Empirical evidence in the history of modern communication confirms this dynamic. Carolyn Marvin superbly summarized the paradoxical relationship between accelerators and brakes, suggesting that it can explain the distance between utopias and reality: "Early uses of technological innovations are essentially conservative because their capacity to create social disequilibrium is intuitively recognized amidst declarations of progress and enthusiasm for the new" (Marvin, 1988, p. 235).

Social Construction of Technology

The concept of Social Shaping of Technology (SST) groups several sociological traditions[4] that challenge the common idea that scientists and engineers work on their inventions completely isolated from society (Bijker, 1995, p. 241). Bijker and Pinch (1987), the main proponents of the social construction of technology, argued that the main consequence of this approach is the ability to understand and explain why the same technology can be shaped into different uses and features in different social contexts. Authors argue that technologies are inherently open to *interpretative flexibility* and *relevant social groups* within societies engage in a power struggle, proposing technological solutions and pushing to impose them. The development of a technology cannot be linear or predefined: there are several possibilities and the one that succeeds is usually the proposal of the most powerful social groups inside society, when it achieves *rhetorical closure*. This approach reveals not only that technological solutions are the result of social negotiations and conflicts, but also that "the actual technical working of the system" is socially constructed (Pinch, 1996, p. 27).

In the evolution of the discipline, the initial focus on the design stage of the technology was extended to the diffusion and social adoption processes (Mackay & Gillespie, 1992, pp. 698–705). Once stability has been reached, a "technological frame" is formed and new definition proposals will be strongly influenced by the adopted frame (Bijker, 1995, p. 252). In this context, the initial *interpretative flexibility* is reduced, but this does not mean that social groups can raise new alternative meanings for a technology and open up new conflicts. What the social construction of technology emphasizes is that in every stage of technological change there are alternative definitions and paths available, and we would only understand the ones that prevail if we analyze why and how they succeeded over the ones that were left behind (Pinch, 2001, p. 397).

This perspective has a more solid research program than most historical approaches (Pinch, 1996, pp. 28–29). Despite the similarities with Winston's position, the sociological proposal is more flexible and invites the researcher to detect the actors that participate in the process of adoption of the technology, map tensions and proposals surrounding it and follow their evolution. A media company can be understood as a social system inside which we are able to identify *relevant social groups*: departments and even individuals that participate in the decision-making process about the use of the Internet as a news medium. Inside a media company, consensus could be reached or a solution could be imposed by the management. Power relations inside the company will obviously influence the results. Nevertheless, it should be taken into account that companies work in a competitive context and are not isolated systems: they constantly benchmark what the competitors are developing. We might think of companies as the societies Bijker and Pinch analyze, which actively import and export technologies that are adapted to the most convenient solution for the needs of the receiving society.

Actor-Network Theory

Taking Social Shaping of Technology to the field of anthropology with the help of semiotics, the Actor-Network Theory (ANT) offers a compelling framework to explain the dynamics of innovation (Callon, 1987; Latour, 2005). For ANT, every element (persons, institutions, material artifacts) related to a technological innovation is an *actor* in the process of defining it: while human actors propose definitions of a technology, material actors may limit the spectrum of possible definitions with their own material limitations. Actors are part of a *network* of relationships that shape the innovation. Inventors of the technology embed some expectations in the design of the artifact, i.e., there is an *inscription* (Akrich & Latour, 1992) of potential users, uses and, rules. This may direct the actual uses, but the network of actors involved in the development of the technology is often complex and full of tensions. In the process of social adoption of the innovation, each actor in the network engages in an active but usually uncon-scious exercise of simplification of the whole structure of relations and compet-ing definitions in order to be able to integrate the innovation "into the context of their specific work tasks and situations" (Monteiro, 2000, p. 77). This process of adapting the definitions (uses, expectations) of a technology to the own needs of each actor is what ANT labels as *translation.*

Latour (1993) insists on the need to use actor categories and definitions when describing a technology, suggesting this is the best way to understand why and how they use it. The research outcomes of ANT have been criticized for overstating the power of individuals and local decisions in shaping a technology, neglecting the broader social context and existing structures of power and interests (Williams & Edge, 1996, pp. 889–890), but the network approach can integrate these macro-structural factors into the analysis without many compli-cations.

Social construction of technology studies have usually focused on recon-structing the history of an already stabilized or rejected technology with some decades of perspective. The main methods have been "detailed reading of historical archives" (Boczkowski, 2004c, p. 147) and interviews with the actors involved in the invention and adoption of the artifact (MacKenzie, 1996, p. 263), with a qualitative approach that attempts to "recover the sociotechnical frameworks within which the actors worked…and look at the world through their eyes" (Pinch, 2001, p. 396). When dealing with current developments, which is more common in ANT, ethnography is usually chosen to address the research objectives (Lievrouw, 2002, p. 132).

Anthropology of Technology

From an anthropological perspective, which has cultural diversity at the core of the discipline, variations in the use of a technology in different societies are as logical as they are attractive. Lemmonier (1993, pp. 6–9) states that the adoption of a technology is a process of selection of technical features for an invented artifact or one imported from another social group. In this process, social actors decide the uses of technology, the working routines and the roles during its usage. These are a set of elections to be made among different open options. Lemmonier labels this process *technological choice.* His main hypothesis is that these decisions are not usually the result of a rational analysis of all the features of the technology, but they are based on other factors: the symbolic connotations of the invention, the commonly defined uses and its relationship to other technologies of a concrete production process. Each society will use a technology in a particular way, because its symbolic context is different.

"All techniques are thus simultaneously embedded in and partly a result of non-technical considerations" (Lemmonier, 1993, p. 4). Some technical aspects of an invention may resist social shaping quite solidly, but different societies can attach different symbolic meanings to a technology. Artifacts may be the same, but their use may have different social implications. Any sociotechnical system is framed in a symbolic and social system. Both need to be studied to understand them. Individuals learn how to interpret the world by assuming the symbolic system of their society. They also learn the socially defined routines for the use of a technology.

This anthropologist focuses on technologies linked to production processes, and argues that inventions have to be analyzed in the context of their daily use. It is in this way that the researcher is able to grasp the uses and meanings of a technology and compare differences between places and people. Artifacts analyzed abstractly may seem homogeneous; it is the daily use that makes the difference and allows the discovery of plausible explanations for the variations found. In online journalism, this would suggest that observing the work in newsrooms can be a very rich experience in order to interpret keys to identify different (or similar) ways to use the Internet. Lemmonier (1993, p. 8) defends ethnography as the best methodological approach to study these phenomena, because only direct observation of the way users manipulate the artifacts in a real production setting can help to understand the meaning that a given technology has got in that context.

Discussion: A Constructivist Tool Kit for Online Journalism Research

The lessons taken from different disciplines and researchers that have analyzed online newsrooms, technological innovation or both, allow us to create a rich analytical framework to understand change and development in the new medium.

Even though some of the disciplines described above have a macro-social perspective, their conceptual framework can be easily adapted to analyze smaller social contexts such as media companies. The sociology of news production is the adequate setting to do this move. Any media company is an organization with certain routines and values based on corporate objectives (Shoemaker & Reese, 1995), the representation of their audiences (Boczkowski, 2004b, p. 175) and the professional culture of journalism (Schudson, 2003). Altheide and Snow (1979) argue that all the previous factors are entangled in the *format*, the shape of the product, including its structure, features, and content. The media logic that shapes this format is partly a product of the technological choices, but overall is the result of corporate and professional strategic decisions, from the objectives and organization of the company to the framing of the audience and the definition of working routines.

These different facets of a media company do interact and mutually influence each other, but also relate to external social institutions and actors, such as other media companies and technology developers and users. The evolution of each element and institution is a change that may affect (and promote change in) the rest of the system. Therefore, the context of the media companies has to be taken into account, particularly:

a) Alternative definitions and uses of the technology by other actors (inventors, users). Mass media use of the Internet was preceded by many other users who proposed specific definitions of the technology and its uses (Abbate, 1999) that influenced the ideal models of online journalism.

b) Traditional media have been maturing a news production model for decades and dealt with technological innovation before the Internet appeared (Boczkowski, 2004a). This general context has to be taken into account, for it may have influenced the adoption of the new technology.

c) Competitors and other media using the Internet are benchmarked when companies take their decisions.

Having this broad context in mind, innovation research theories point out which are the focal points to analyze the process of adoption and the shaping of the *technological frame* of online news (understood not only as the technical features of online journalism, but also its definition, working routines and user skills) inside media companies (understood as actor networks). These would include:

• Comparing variations in the uses and definitions of technology between newsrooms. Identifying the *translations* of the ideal models of online journalism into daily routines in the production culture of online newsrooms.

- Detecting relevant actors (social groups, artifacts, institutions) in the decision-making process for the adoption of the Internet in the newsrooms, documenting the power relationships and the competing (or not) definitions.

- Checking if each actor (again, not only people, but also other elements such as technical artifacts) can be labeled as accelerators or brakes, by interpreting their needs and fears. The researcher should identify ambiguous positions and assess change over time.

- Identifying actors' strategies: some may use verbal or written proposals, others with less power would prefer daily activism, by altering proposed work routines.

- Finding the *rhetorical closures* that lead to consensus imposed by some of the actors and assumed by the rest. These should be the basis for the current technological frame and define the main path for further evolution.

- Locating technological choices in the process and differentiating elements of the technology more solidly opposed to social adaptation, and others that have been changed to fit the newsroom logic.

- Confirming that the newsrooms have reached the point where choices are assumed as natural.

At the present stage of development of online media projects, when they are already established and operating, research must focus on the newsroom, where technology is being used and definitions are applied and, maybe, disputed. The researcher needs to be aware that the top management levels of media companies were usually involved in the initial steps of the adoption of the technology and that the business logic has played a role besides the journalistic logic that is more visible in daily newsroom routines. Furthermore, the offline newsroom in traditional media companies with online projects, and the technical department, are actors closely related to the activities of the online staff. Exploring all these actors thoroughly may be impossible for a single researcher's project, but they can be considered from the perspective of the online newsroom, the main *locus* for the development of online journalism.

Some authors have stressed that constructivist research is useful to empower social actors to further develop the technologies they use (Bijker, 1995, p. 253; Pinch, 1996, pp. 34–35). The proposed theoretical approach makes the path of technology visible, as well as the taken for granted decisions and the structures built during the adoption of the Internet. The results of research inspired by this framework can help online journalists to have greater control over what they want to do with the tools they have. "There is no one inevitable logic of development. There is choice" (Pinch, 1996, p. 34).

NOTES

1. In fact, this book is the first edited collection of studies of this third wave, and the editors hope it will encourage much needed online media research based on a nondeterministic theoretical framework.

2. *New media research* has been used widely to label the research on communication and technological innovations since the late 1970s and especially in the 1990s, particularly to refer to Internet studies (Lievrouw et al., 2001).

3. A more systematic definition of information and communication technologies is provided by Lievrouw et al.: "Technology includes the artifacts or devices that enable and extend our abilities to communicate, the communication activities or practices in which we engage when developing and using those devices, and the social arrangements or organizations that form around those practices and devices" (2001, p. 272).

4. Williams and Edge (1996) and Lievrouw (2006) offer a contextualized panorama of social construction of technology and actor-network theory, the two main currents in SST.

CHAPTER TWO

Ethnographic Media Production Research in a Digital Environment

Roel Puijk

Puijk examines the methodological challenges and opportunities delivered in by technological and organizational change in media newsrooms. He uses his own research in the 1980s and the 2000s to highlight the differences in access negotiation, data gathering and internal and external communication processes. As journalists do more and more tasks within computer systems, the researcher needs to find ways to understand what is going on in the digital networks beyond the physicality of the newsroom and to take advantage of the wealth of data that is stored and exchanged there.

EDITORS' NOTE

Media organizations have changed radically the last decennium. Increased competition and technological developments have given an impetus toward new production modes, changes in organizational structures and ways of thinking about the readers and viewers. Terms like digital production, multi-platform production, media centers, and interactivity indicate some of the factors involved. The changes have not been unnoticed in media research and a renewed interest in ethnographic studies of media production, of both news and other genres, can be found (Cottle, 2000, 2003).

When reading a master's thesis in informatics about multimedia production in the Department of Culture of the Norwegian Broadcasting Corporation (NRK), I was struck that the rather technical evaluation of how software is used revealed interesting information about the relationships between the journalists in the editorial units under consideration. The study (Dalberg, 2001) shows how the same software was used differently in radio and television. The radio journalists used the software as a common archive and could access each other's texts, even before they were finished. This facilitated reuse of the same material on different platforms, but sometimes unfinished texts were being used on the website, causing tension between the journalists. The television production unit used the same software as a broadcasting tool. This functionality conflicted with using the software as archive, so the texts (commentaries) were either not written down or were archived locally. This inhibited the reuse of the material on the web as the online journalists had to rewrite the reports after they were broadcast. Again, there was a basis for conflict between the different journalists about the way they used their computers and software. Despite the unfamiliar terminology in the thesis, it was interesting to learn about the social use of software, as it involved patterns of cooperation and conflict between journalists working for different platforms. As a media ethnographer it also made it clear to me that we have to take the possibilities and use of software and the digital working culture around it into consideration when studying editorial processes.

During the 1970s and 1980s a number of media sociologists went to study news production in situ—observing journalists at work (see Tuchman, 2002 for an overview). Not all of these studies, that Cottle (2000) calls the first wave of news ethnography, reported in depth on the methods they used, but it seems that they took their pencils, notebooks, and sometimes tape recorders to the media organizations, clipped stories from the newspapers and taped the TV programs they studied. At least that is what I did when I did fieldwork in NRK in the 1980s. Here I studied the production of nonfiction programs (Puijk, 1990). In 2003, I went back to the same corporation to study how they adjusted to the new digital situation. I will use these experiences to reflect on what consequences the changed media situation, including the introduction of computers and Internet-based communication, has on the research methods of media production ethnography. Both communication patterns and the amount of information that is involved in the research have changed, sometimes limiting the researcher's insight and sometimes opening up the inquiry. Doing ethnographic research of media institutions today will often imply that the researcher not only has to be aware of how data mediated communication functions when positioning him or herself in the organization's information flows, but that also ethnographic research can profit from data communication techniques and software tools.

In this chapter,[1] I will deal with access to information from the point of view of ethnographic research. This is a question of negotiations between researcher and gatekeepers at different levels—it involves not only the crude question of access to the field, but also more subtle forms of access to meetings, interviews, archives and, lately, also to material that circulates in computerized systems. This consists mostly of text-based material generated not only within the media institution but also in their relations with sources and audiences. Before going into detail of what kind of information may be considered, let me start with a short description of the changing contexts in which media organizations operate today.

Changing Contexts

During the last 25 years, media organizations have changed in many ways. In general, competition between media organizations has increased. This applies to both commercial media and public service broadcasters. Commercial media tend to have been integrated into large media conglomerates that are not bounded to national boundaries (see McPhail, 2006). Although not integrated in (transnational) media conglomerates, Western European public service broadcasters also have undergone severe changes. From being mono- or duopolies they have become involved in competitive markets where they face commercial actors (Lowe & Hujanen, 2004; Lowe & Jauert, 2005). Increased competition, combined with changes in (media) technology, has resulted in, amongst others, increased pressure on productivity, changes in divisions of labor and frequent reorganizations. Also, the Norwegian Broadcasting Corporation has had its share of change in the form of legal transformation and reorganizations.

Competition and production pressure is not the only factor in frequent reorganizations. Closely connected to these are the increasing technological changes. The introduction of the Internet is, of course, a major factor here. Most media organizations have entered this field and produce content for the Internet in addition to traditional media output. They are also experimenting with other techniques and platforms (podcasting, streaming video, mobile phones, PDAs, etc.), in fact turning into multi-platform production houses. In NRK, multi-platform production was one of the main factors in a reorganizing process that started in 2000, when radio and television departments fused and were extended with Internet production. This of course augmented the complexity of production with consequences for the role of the researcher—in terms of amount of textual material to overview (and archive) and possible foci of research.

The entrance of computers in media organizations is not only confined to new production outlets, but has also changed production processes, communication patterns and labor routines. In media research, more focus has been on the consequences of digitization of the production chain for the division of labor and the resulting convergence and divergence processes. From an ethnographic researcher's point of view the changes in the workplace are relevant as well. The introduction of new communication channels (computers, email, intranets) in the office space changes both the observer and journalists' access to information. These changes are so common also outside media organizations that they almost are unnoticed, but they have consequences for research—not only because those observed have obtained new tools of information sharing, but also the observers are equipped with similar tools. Before assessing some of the implications of these changes, I will first look at the negotiation process that is the deciding factor for doing fieldwork at all—permission to do research.

Gates and Gatekeepers

Entering NRK's television building in the 1980s was easy—as long as one entered with a firm determination and did not hesitate when entering the lobby, the person at the television desk would not bother and try to stop you. Today, one has to be equipped with an identity card or to have an appointment with someone inside who has to meet you at the entrance. But entering the building is just one aspect in the negotiation of access—a process involving not only the official gatekeepers, but also relations with the people that actually are observed.

As Lindlof notes (1995, p. 110) it is not always obvious who is the gatekeeper—i.e., who has the formal power to accredit the researcher's access to the organization. Entrance and access are often not a question of one person controlling the gate, but the result of internal negotiations involving different parts of the organization (Garcia, 2004).

When preparing for my first research visit I started at the ground level. It took the department several months to decide. When I finally got permission, it was based on the condition that I, after a short stay, would limit and specify my research proposal.

In 2003, I wanted to revisit NRK to see what had changed after all these years. In particular I wanted to see how new technology influenced the production process. This time I did not start at the "bottom" but at the top. I wrote an email to the director of NRK-Oslo and asked permission to do fieldwork in the organization. The initial response came in an email:

We have discussed your request in the NRK-direction and our answer is as follows: As NRK is a big and important institution in Norway we want to be open for research projects. But we are skeptical to open up for the kind of 'observations' you sketch. We are afraid that this will disturb the work of the production teams and that it will influence having outsiders being present in this way. We are though positive to put relevant interview objects at disposal and possibly written documents according to further agreements.

Of course, this was not a very encouraging answer but one phone call and a meeting with the director later, I was invited to a meeting with the head of the Department of Factual Programming and the project leader (editor) of a culture talk show. It is difficult to know exactly what arguments were responsible for the change in attitude—whether it was my earlier book that I brought along to show the seriousness of my request, whether it was my position as an associate professor and the argument that we needed this kind of research to be able to educate television professionals adequately or whether is was just that personal contact made the whole project less threatening. Probably it was a combination. In addition, I accepted to let them make the choice of production units I would follow.[2]

Once I had access to two production units, I was provided an identity card and had physical access to most areas of NRK. I was free not only to participate in the daily activities of the production teams (i.e., staff meetings, recordings), but also to interview people from other departments. This is not to say that I had free access to all of the information I wanted and that there were no conditions. I will come back to this later.

If we compare the two access negotiation processes, we can note that the top level of the organization has much stricter control today. The process of redefining the relations with their environment implied a diminishing of power on the ground level and an increase of power at the management level. The administration of identity cards at a central level is one factor that helps the top management in this respect.

Working Environment and Communication Routines

The reorganization of NRK in the early 2000s resulted not only in new structures on the organizational map, but also physically many changes were made— new production units have been established, they were moved across the buildings, etc. The production units I followed in the 1980s and in 2003 were situated in the same building and even in the same corridor. But the physical outlook had changed: instead of entering a corridor with offices on each side, the production office now was an open space used both as office and as studio. The journalists and technical staff were grouped and sat around four working stations with four computers each. They shared an editing suite and control unit with another production unit, situated in between the two open office spaces. In one corner, there was a meeting table where I could sit with my laptop.[3] This situation in many ways was advantageous for observation—it was easy to oversee who spoke to whom and what was going on. Sometimes some of the members would gather around the meeting table to discuss the details of a program. Compared to the former situation where I often felt I had a "private" office, observation and talks about things that happened felt more natural for me in 2003.

Even though the new situation was advantageous for observation, I also experienced that other aspects of field research were more difficult.[4] They have to do with digitization and access to information, resulting on one hand in restrictions on access and on the other hand in problems of abundance of information.

Internal Communicative Space

The introduction of computers in media organizations like NRK consists not only of sound and video production chains being converted from analogue to digital. The introduction of computers has also changed communication and information patterns within the organization. This digitization process including the use of computers in administration, intra- and Internet and other computerized ways to deal with internal information for coordination of the members of the organization is not restricted to media organizations but is a feature that these organizations share with other organizations. Nonetheless, this doesn't mean that we should not take it into account.

Internal communication is a central element in organizations. In order to get people to work together, to adjust parts of the organization to each other and to create a sense of identity, information has to be shared among the members of the organization. Every organization has its own way of organizing the information flow; they arrange regular and ad hoc meetings, forms are completed and supply central coordinating departments with input, while generated data are again distributed to the appropriate members. Of course, the flow of information is not evenly distributed—some information is shared among all members of the organization, while other information is shared within groups or between individual members. Every organization develops its own way of dealing with this information flow, deciding what channels to use (one-on-one meetings; memorandums that are distributed among certain members of the organization; forms to be filled out; internal newsletters, etc.) and institutionalizing new routines when flaws are detected. Although there may be individual differences, the members of the organization are positioned in this internal communicative space almost automatically according to their positions. Also, the researcher is positioned and positions him or herself, but this is a more active process, involving negotiations as to what information may be (made) available. Gaining physical access to observe is often only one step — during my fieldwork I often had to negotiate not only whether I could attend specific meetings, but also whether I could get copies of minutes and other relevant written material.

While much information formerly was exchanged in meetings, internal bulletins, copied documents, etc., now much is transferred to data based systems (intranet, email, etc.). The existence of electronic internal communication systems has several consequences for ethnographic fieldwork. If the researcher is not connected to this system, which might also contain what is considered sensitive information, he/she will not being able to follow the discussions and will have to rely on verbal information from one of the team members. I did not ask to be put on the internal email list and was not connected to NRK's intranet and sometimes found myself outside the communicative space the members of the production unit were part of. Sometimes this resulted in showing up at the wrong time and not being informed of the last-minute changes that had occurred.

We all know that the material that circulates in these systems vary—from factual information to entertainment and emotional outbursts people send to each other. It is impossible in one chapter to give a detailed overview of how this affects relations within the organization, however, let me give an example of how these channels were used to give feedback: I was talking to the online editor of a production team. While talking he also kept an eye on his computer; suddenly he got a smile on his face and explained that he had received an email from the main Internet desk (NRK.no). They had published his article on the first page and thanked him for good "desking." The next day the leader of the production unit also mentioned this during the meeting of the production staff.

NRK is a rather large organization and the buildings are complex so people who collaborate do not necessarily meet in person very often. Even though people might see each other during lunch in the canteen, members from the same department tend to sit together and talk. Of course, several forms of feedback are used, but when it is in the form of an email as in this case, it is very easy to distribute this to others and it then it used to draw the attention of more members of the production unit. This can be used strategically to gain visibility in an environment that is characterized by flexibility and change—both for freelancers and other staffers who have to prove themselves much more today than before. At the same time, using email and SMS (cell phone text messaging) to give simultaneous feedback is also part of the culture of a younger generation. Not only did I note that colleagues, but also close friends and family sent SMS during and after broadcasts to program hosts and project leaders, mostly with a (sometimes funny) remark on the program. Sometimes these SMS were read aloud on the production site.[5]

Data-Based Production

Data-based systems are used in journalistic production. Internally a new division of labor follows digitization. Reporters now log their raw material, pre-edit, and work for different platforms. Interesting questions also arise from how journalists use their computers not only to share but also to hide information from their colleagues strategically. The study (Dalberg, 2001) mentioned in the introduction of this chapter shows the tensions between sharing information about items according to institutional ideology of multi-platform production and the individual journalist's interest to keep some items for themselves to be able to present them as "their" story (scoop) in order to strengthen their career (Erdal, 2007). Realizing that these kinds of tensions exist within the organization it becomes important to know how these data programs function and how they can be manipulated. This, of course, demands competence in the field of production-software by the part of the researcher.

Several studies have shown that digital competence and attitudes towards using computers varies among the employees of media organizations (Rintala, 2005; Engebritsen, 2007). This implies that we have to be aware of the fact that not everyone is very confident with using all the possibilities the software provides for and that people make their own standardized routines. But it is not only a question of competence—in media production the question of time constraints is ever present. Knowing and judging how much time operations take, is part of the working routine and may explain why certain options and shortcuts are chosen instead of more time consuming ones.

In these and other cases it may be important for the ethnographer to understand how data programs function—my understanding of the working process and the functioning of publishing tools increased dramatically when I edited my own Internet site based on similar software.

External Relations: Sources, Viewers

Data-based communication is not only important internally, but has also changed the relation between journalists and their sources. Tuchman (1978) uses the term "news net" in her analysis of the routines of journalistic information gathering. One of the routines she describes is that of beat reporters that establish routines by regularly surveying hospitals, police stations, and town halls, etc. Today, journalists often use the Internet to look out for potential stories.

In his study of Internet journalists in Catalan media, Domingo (2004, 2006; see chapter 7 in this book) found that they relied almost entirely on electronic communication (from news agencies) for the production of online news. In my case of nonfiction programming the journalists often browsed the Web for ideas and themes they could use for the program. Several systematically researched specific Internet sites (e.g., governmental and NGO sites) in their search for potential stories. Therefore, researchers should ask for a copy of the bookmarks journalists have in their browsers or to be allowed to see their browsing history, to have very rich data about their online sources.[6]

Journalists do not only generate themes and stories, but are also contacted by their sources, as they want to promote a theme or a specific case. In the latter instance, the Internet is frequently used and access to the unit's email can provide data on what ideas are proposed. Ytreberg (2003) reports that he had access to the emails received by the production unit he studied, getting insights of the general communications directed to the unit. Sources have often built relationships with specific journalists and approach them directly.

With the arrival of Internet and digital interactivity media organizations are much more prone to gather feedback from their audiences. Of course letters and telephones were used before, but digital communication enables direct interaction with members of the audience. This has become a regular and organized feature. A wide array of services have been developed to enable the viewers to communicate with the organization, ranging from chats, text-messaging, online polls, call-ins, to publication of email addresses where people can communicate directly with members of the production units. A large portion of this user-generated content is public, as it is posted on the Internet pages of the media institutions.

In other cases, input from the audience is used more directly in the program. Here the communication is less public. One of the television programs I studied ended their program with a section where a health expert was interviewed by the host of the show. After the TV broadcast, the interview continued for thirty minutes and was streamed live on the Internet. Viewers could write their questions on the program's website and their input was received during the webcast. A moderator picked out the questions she thought were the best, edited them and sent them to the program host's computer so she could read the questions word-for-word from the screen. I was allowed to have the file with the original questions as they were written by those who posed them, which allowed me to analyze the whole gatekeeping and rephrasing process, the changes that were made from "raw" questions, formulated by the viewers themselves, to the questions asked by the presenter, based on viewers questions but reformulated by one of the production staff.

Multiple Texts

In the 1980s the output of the production process was rather easy to collect and archive. The television programs produced could be recorded and stored away on videotape. Today the situation has changed dramatically, first of all, because several media are involved, and in particular because Internet is used as output media. One of the main characteristics of the Internet is that it is a dynamic medium—not only can it be updated at any time, but the output can also be different according to specific characteristics of the user (pages generated "on the fly"—see Manovich, 2001). This implies that what was clearly delimitated and fixed output before, now has become variable, a flux. For a researcher it poses several problems. One of these is that a vast amount of material is produced and should be overlooked and archived. When doing fieldwork it is often unclear which parts of the published material will become important for the analysis. An additional problem is that with cross media production different platforms are used in relation to each other. This implies that it is important to document the time of production and publication (Puijk, 2007).

The flexibility of the Internet complicates the aim of archiving content (Brügger, 2005). Pages that were available one day may have disappeared the next. Some of the media institutions in Domingo's study regarded the news compiled by their online unit as "provisional" and replaced it by "real news" when the printed version of the paper was available (Domingo, 2006, p. 379). In case of less news-oriented material like the Internet site connected to television programs I studied, material might be available later, even though some of the minor changes get lost.

Internet sites in professional organizations are normally built using databases, generating pages when the user requests them. Some of the material may be available and traceable if one knows the principles of generating Internet pages. However, one must always be prepared that Internet pages and television programs that are streamed on the Web might disappear suddenly (changes in policy, copyright issues, etc.). Having good routines in one's own dealing with Internet output is essential.[7]

Limits on Information

As media organizations have become more competitive during the last decade, so too have public service organizations like the NRK. They now compete in a market with national and international commercial actors. In many countries, this has resulted in demands for a more suitable, flexible organizational foundation. In Norway, NRK was a state institution until 1986. It became a foundation in an effort to compete in the new liberalized audio-visual market. From 1996, the status was changed again to a state-owned limited company. These changes also have a direct impact for research. For example, the fact that the Norwegian Freedom of Information Act is not applicable anymore and information about the organization may be kept secret from a researcher. This makes it more difficult to get information about the relation between NRK and its owners (formally the state represented by the Minister of Culture). They argue that to give certain kinds of information could undermine their position in the market. This was the reason that they restricted the information I could obtain on economic matters. During the course of my fieldwork, there were several instances where I had to ask for information that was on the border of public and private. In these cases, the editor would ask her superiors whether she could provide the information to me.

In other cases, the question was more problematic. Let me give an example: In the course of my fieldwork, I attended meetings where the members of the production team talked about their budget. Because of internal budgeting, the production team rents its technical equipment from the technical department. It turned out that the amount they had to pay for renting the newly installed digital production chain was much higher the second year than the year before. I thought this was interesting information because it might indicate that this chain was underpriced the first year. This question could be taken further: Was this a conscious move from the technical department, trying to push new technology? This question is very relevant to my study, but it will be difficult to answer without the help of the economic figures as documentation.

Conclusion

Doing ethnographic research will always focus on observations and person-to-person contact with the actors in the field. In this way, doing ethnographic research has not changed fundamentally in the last decades. Still the introduction of computers has both restricted and widened the amount of information accessible to the ethnographic researcher. Depending on what access one can negotiate, the information flow inside computers can be hidden or it can provide a rich source of information. In many cases (e.g., email), this information is text based, which has the advantage of being fixed so one can retrieve it later. But the information has to be contextualized and interpreted with care. Email communication, for example, in many ways lies between oral and traditional textual utterances. Its status of having oral traits (often formulated in a hurry and in a personal fashion), not being as "official" as other written texts like memos and minutes, has to be taken into consideration.

Doing research in a digital media environment implies more and more often that the researcher is familiar with computing at a more advanced level. Sometimes it implies insight into the computer programs journalists use, sometimes it may be a question of using tools to analyze aspects of Internet production. This has to be taken seriously when preparing for fieldwork. Of course, the techniques and tools used will depend on the research questions posed.

NOTES

1.   This chapter is a result of the research project Television in a Digital Environment.

2.   Of course one has to reflect on how this influences the research (Puijk, 2004).

3.   The table would be removed when they used the same space for recordings on Monday—the program shows the work environment in the studio-based sequences.

4.   See also Domingo (2003) for useful advise on how to deal with the fieldwork situation, based on his own fieldwork in four Spanish online newsrooms.

5.   My observations indicate that this practice may be particular widespread among the female leaders and their close relations, but as both project leader and host in the editorial units I observed were female, it is difficult to draw a conclusion.

6.   Software to compile lists of visited Internet sites is available, and the information can be anonymously transferred to the researcher (www.attentiontrust.org/services, www. attentionbank.com).

7.   Sometimes backup archives are available, having backups of the site at particular dates (www.archive.org). Although useful, in my experience they have their limits, as they relatively often store just a small part of the elements of the website, or sometimes just don't work.

PART TWO

New Media, New Routines?

CHAPTER THREE

News Production in an Irish Online Newsroom: Practice, Process, and Culture

Anthony Cawley

In Ireland as in much of the world, major city daily newspapers moved with initial caution into the world of online journalism in the mid- and late 1990s. Since 2000, few medium- or large-sized newspapers were without an online division, but in the early part of the decade the online operations of established papers around the world continued to struggle to find their own identity. Cawley's ethnographic narrative from one such newsroom exemplifies the "thick description" anthropologist Clifford Geertz (1973) advocated, and reminds us of the rich sociological detail offered by Herbert Gans and other early researchers of news production.

EDITORS' NOTE

This chapter examines *The Irish Times* and its website, *Ireland.com* (www. ireland.com), as structures to produce and publish news, and looks at the relationship between the old organisation and the new organization, between the old media form and the new media form, and the strengths and conflicts that come from their close and direct relationship.

After an introductory interview with the website's deputy editor in June 2001, I was granted a three-week research placement in the online newsroom during August of the same year. It wasn't quite access to all areas, but I was allowed to interview whomever was happy to be interviewed. I captured opinions from across the online operation (editorial, technical, marketing, advertising) and tried to relate the questions I asked to the professional practice I had witnessed. The following sections, unless flagged otherwise, apply directly to those three weeks of ethnographic study.

The Newsday

"Can't do color." Journalist Paul Anderson frowns at the computer screen: deadlines don't allow for luxuries such as writer's block. He begins tapping on the keyboard and words pop up on the screen as the intro gathers pace. He pauses mid-sentence, then holds down the delete key, hauling the cursor back over what he has written.

The Irish lottery hasn't been won in weeks. The draw is tonight. On a slow newsday, a jackpot of several million Irish pounds is worth a couple of hundred words. He decides the lead will have to come later and he jots down some information to prop up the story: how many weeks since the lottery has been won, the value of the jackpot, statistics on the six numbers that are drawn most often.

Words exchanged with colleagues at the next desk prompt a mini-huddle: who's in the office lottery syndicate, who isn't, who has paid their share this week, who hasn't. The conversation dissolves before it comes within earshot of editors who mightn't appreciate such distractions in the newsroom late in the afternoon.

Another burst of typing on the intro; another reach for the delete key.

"Can't do color," he says.

Words come easier at the beginning of his shift, when he is working on a hard-news story. Assemble facts, gather quotes, structure them in order of importance. An Irishman has been arrested during protests at the G8 summit in Genoa, Italy: the local hook on an international story. Anderson's first run at the story lays down the bare facts: the man's name, age, where he was arrested, how long he has been detained. For his second run, he consults his contacts book for the number of a source within the Irish anti-globalization group to which the arrested man belongs. The group is vocal more than militant, and Anderson's source is willing to talk. His questions flesh out the facts. Have you spoken to the arrested man since he was detained? How have the Italian police treated him? What is his state of mind?

Anderson transcribes each answer carefully into his notebook. The Italian police are deporting the man to Ireland this evening, the source tells him. He revises and updates the story on his computer screen.

His next run at the story involves a call to the Irish embassy in Italy. The embassy confirms that the man is being deported this evening: the flight number or time isn't known, but it is a direct flight. He phones Dublin Airport to check flight times from Genoa this evening. Only one flight is scheduled. He works the airline name and expected arrival time into the text. Unless something unexpected breaks, the story cannot move forward until the flight arrives. He opens a fresh page of his notebook and lists people and organizations he should phone for his next story.

The newsroom looks like a normal print newsroom: messy desks, coffee mugs, background noise of ringing telephones, computer screens displaying stories in various stages of completion. The journalists are a bit young and casually dressed, but little else stands out. On Anderson's desk are the normal tools of a print journalist: a notebook, a pen, a telephone, his contacts book, a telephone book, a tape recorder, and a computer. His work practices resemble those of a print journalist: he calls sources, press offices, organizations, writes down what they say and assembles the story into an inverted pyramid structure.

One difference is significant, however. Each time he finishes a run at a story—adds fresh information or reaction—he publishes the update directly himself. A reader who has been paying close attention this morning would have seen the G8 arrest story evolve from bare facts, to having reaction from a source close to the arrested man, to having official confirmation of his deportation, to having his flight and expected arrival time. Anderson controls the information gathering process, the writing, the editing and the publication of his story. He controls stages where, conventionally, an editor, a page setter or a printer would have assumed responsibility on the story's journey from the newsroom to the public domain.

This is an online newsroom, Anderson is an online journalist, and the traditional demarcation of news production doesn't apply. He is a senior journalist (in his late-thirties and older than the others) and is allowed responsibility to publish directly to the website's breaking-news section, without prior approval from the sub-editors or news-editor. Not all the online journalists enjoy the same privilege, although most of them can still update their stories during the day as they gather additional information or events break.

The ability to publish immediately brings pressure to publish frequently. The deputy editor calls across the newsroom. The breaking-news section hasn't been updated with a fresh story for forty minutes. He wants copy.

Despite the constant imperative for something new, the attitude is far from publish (right now!) and be damned. Immediate publication risks immediate libel, especially as online news production bypasses some of the layers that can filter out libel in a newspaper. Anderson and everyone else with the status to publish online are directed to an internal memo: "Nothing should go live without being checked independently" (internal memo, August 2001). The website has yet to be sued, and its editors don't want to test the company's commitment to the online operation with a heavy financial penalty in court.

"Can't do color," repeats Anderson.

He has chiseled a few hundred words that are acceptable for the lottery story: the intro ends up being a play on the six numbers that are drawn most often. He holds off publishing it online, deciding that the piece needs a photograph. A visit to the technical department, two desk rows away, he secures one of the digital cameras. I follow him to the lift—the newsroom is on the top floor, and going down four flights of stairs can be an intimidating thought to a journalist late in the afternoon. Heavy rain is spraying Dublin's streets, a circumstance that is factored into an important editorial decision: where to locate the photograph? Destination, the nearest newsstand with a lottery machine. The online newsroom is in Dublin city center and the nearest newsstand is three doors down.

A small queue stands by the lottery machine. Anderson explains who he is and what he wants to the shop assistant and the woman at the head of the queue. The woman agrees immediately, the shop assistant hesitantly. He takes the photograph: the shop assistant stands beaming behind the lottery machine; only the woman's hand, holding a lottery ticket, makes the shot.

Back to the lift—going up four flights of stairs is an even more intimidating thought at this time of the day. The photograph is loaded onto the network. He resizes it, attaches it to the story, and adds a caption. He loads the story online.

Today, he has been journalist, sub-editor, photographer, and publisher, crossing boundaries of demarcation that would not be tolerated at a print newsroom. The boundaries have long since blurred for him: he expects to apply a range of skills to news production in an online medium, and he expects to report on a wide range of subjects. Apart from sports and business, all the online journalists are general reporters.

"We're not in a position to have correspondents. Not for a few years anyway," he says (interview, August 8, 2001).

He returns to hard news; his mood is lighter already.

The History

The Irish Times was established in Dublin in 1859 as a conservative broadsheet newspaper. In the years before Irish independence from Britain in 1922, and in the decades after, the newspaper's loyalty favored the government in London over the one in Dublin. This marginalized its appeal to a small, select section of Irish society, so much so that a former editor quipped, with only mild exaggeration, "It used to be said that every time a name appeared in the deaths' column of *The Irish Times*, the circulation of the paper went down by one" (Oram, 1993, p. 51).

The Irish Times began to turn into a liberal newspaper with a broad appeal in the 1960s, reflecting changes in a modernizing Irish society. Today, it holds a reputation for impartial quality journalism in Ireland akin to *The Guardian's* standing in Britain or *The New York Times* in the United States. In 2003, it became the first Irish national daily newspaper to appoint a woman as editor.

The Irish Times launched an online edition in 1994, and was the first Irish newspaper to do so. The decision to set up early, establish and maintain a lead as an online source of Irish news, was based on "a combination of luck and vision, and a lot more luck than vision," explained online deputy editor, Conor Pope (interview, June 14, 2001).

Irish Times executives were traveling in the United States, scouting for a new printing press. One of their visits brought them to *The San Jose Mercury News*, whose editor—as an aside after the meeting had ended—showed them the newspaper's online edition. The executives left without finding a suitable printing press but with the idea that a website could reach Ireland's diaspora and open up a new channel for Irish news.

The website has gone through three distinct development phases between 1994 and 2001:

First phase (mid-1994–1995): The Irish Times Group establishes a website, *The Irish Times on the Web*, to republish the content of the newspaper. Growth in traffic is quick, with about 80% originating outside of Ireland. Few journalistic skills are applied directly to the website. About ten Irish Times stories are reformatted in HTML and posted live. The website has no graphics or photographs. The company has no Internet skills, so computer students from nearby Trinity College are hired to put the website live.

Second phase (late 1995–1999): The company's management is satisfied that the Internet offers the newspaper a channel to reach a new and wider audience. It increases investment in the website, which will be run by a subsidiary company, Itronics Ltd. Four online journalists are contracted to improve the process of repurposing print content to the Internet. A technical team is hired to ensure the website runs smoothly. Photographs and graphics are used, but sparingly to save on download times. Near the end of the phase, a rolling-news section is introduced. For the first time, the online journalists produce their own copy exclusively for the website.

Third phase (late 1999–2001): The company commits additional investment to the website. *The Irish Times on the Web* becomes *Ireland.com*. Rolling-news becomes breaking-news, and the range and volume of stories increases. A larger online editorial team is recruited; it works independently of the newspaper, has a separate newsroom, and produces copy exclusively for the new breaking-news section. The website offers services, such as email and genealogy searches that are far removed from the traditional interests of the newspaper. The number of full-time staff (editorial, technical, marketing) rises to 57. The online operation delivers news content to mobile phones.

Online News Production

In a busy newsroom, the background scenery can be of people with telephone receivers to their ears, and the background noise can be a chorus of, "Hello, my name is X and I'm a journalist with…" In the online newsroom, the sentence ends most often with, "*The Irish Times* website," less often with, "*The Irish Times* online," but least often with, "*Ireland.com*."

In 1999, the website replaced the title *The Irish Times on the Web* with *Ireland.com* to escape being pigeonholed as an online newspaper. A separate brand, so the logic ran, would allow greater freedom to explore areas outside the traditional franchise of the newspaper. The website has, indeed, won territory of its own: a separate editorial team and web-exclusive services have moved it beyond being, simply, a newspaper grafted onto a digital distribution channel. To emphasise that *The Irish Times* is only a part—not the whole—of the website, a link on the homepage invites users to "click here for *The Irish Times*."

Yet, the online journalists in their routine professional practice establish their legitimacy the quickest by referencing their association with the newspaper. Authority is a key issue in any newsroom and has consequences for the production and consumption of news. In production, sources are more likely to divulge information to a reputable media outlet, and for consumption, the public is more likely to accept the reported information as accurate.

In Ireland, *The Irish Times* carries credibility matched by few other news sources. For the online journalists, it is a pragmatic approach to highlight their association. They work in an online medium but resist the hype surrounding it. Many people do not use the Internet; many people who do use the internet don't use it much; the online journalists don't assume the people they call will know what *Ireland.com* is. Reaching for the credibility of *The Irish Times* title seems an efficient journalistic practice and not an abuse of the association, as "online" or "website" is always added to distinguish that the phone call isn't from the newspaper. The distinction is important, also, in the event of an online journalist ringing a source for breaking-news today and a print journalist ringing the same source for the newspaper tomorrow. The duality of news-services provided by *The Irish Times* Group—immediate and on a 24-hour cycle—creates such scenarios.

An ongoing and wider strategic theme is at play here, beyond how online journalists identify themselves during phone calls: the website is trying to escape the company's print traditions (in immediacy of publication, and experiments with audio and video) while also being bound to them (in text, the primary form of communication).

Inheriting the newspaper's writing house-style causes tension in the online newsroom, because it doesn't dovetail with the preferred aesthetics or functionality of the website. The newspaper's house-style diffuses into the website, first of all, through repurposing the print content to the Internet. The print content isn't edited significantly for style or length, and reformatting for the website wouldn't be an unfair description of the repurposing process. Online deputy editor Conor Pope admits that, before breaking-news, when the primary online journalistic task was repurposing, "there was a lot of fairly dull cutting and pasting on behalf of all the editorial staff" (interview, June 14, 2001).

When text is the primary form of communication, writing house-style is a crucial marker of identity. Irish Times senior editorial executives and management insist print house-style be applied to the website to maintain consistency with the newspaper's identity. But writing for the Internet shouldn't be treated the same as writing for print, says Pope. "Very rarely are people going to want to read 30 to 40 paragraphs [online] on a news topic. They'll want to read three or four paragraphs. You want your information fast. The way people read on the Web is very different from the way people read print—it's a lot less leisured.

What you have is people at work, logging on, having a quick glance at the headlines, seeing what's going on, and leaving" (interview, June 14, 2001).

Online sub-editor Jason Michaels tells me that certain conventions of the newspaper's house-style look "anachronistic" and "stilted" on the Internet, such as the requirement to always use a person's full title (Mr, Ms, Mrs, Dr). It may seem a trivial matter, but it grates the online journalists that the formality of the writing can jar the immediacy of breaking-news consumption. Michaels believes that shorter stories are preferable on the Web, no longer than a screen and a half in length, to avoid excessive scrolling by the user. That is the ideal type, although stories will be written longer if they cannot be told in this frame. "We do not want to be a slave to length at the sacrifice of the content," says Micheals (interview, August 16, 2001).

Here, the repurposed print content betrays its origin in another medium: much of *The Irish Times'* news, features, analysis, and opinion spans beyond the ideal length of a screen and a half.

Ireland.com has greater control over the layout and design of web pages. Easy navigation is the prime aim. For the last 18 months, the online editorial staff has been working on a style-book for good practice in online publication. Pope says, "I'd be lying if I said we had a ten point plan for producing perfect web pages…. I don't think we have a perfect template of dos and don'ts for publishing on the Web. It's an entirely evolving process. We're getting better, but we're not there yet" (interview, June 14, 2001).

The Editorial Meetings

On several occasions, without prompting, *Ireland.com* journalists tell me that the "atmosphere" in the online newsroom is different to that in the print newsroom. The comment is aimed less at being negative about the print newsroom and more at emphasizing the positives of their own environment.

On the surface, differences between the two newsrooms seem slight. The physical differences cannot be denied: the online newsroom is smaller, the staff is younger and dressed casually. Atmosphere isn't physical, however, although differences do not spring forth from intangible journalistic attributes: approach, practice, attitude. The journalists, online and print, approach their tasks in a broadly similar way. Attitudes to professional practice aren't inherently different: they share a commitment to quality, balanced, and accurate reporting on issues of public interest.

Despite this, the consensus persists among the online journalists that the atmosphere is different, although they do not articulate clearly how and why. The attitude seems to stem from their self-perception of the dynamic of the online newsroom and their assumptions of the dynamic of the print newsroom.

The online newsroom is in leased space on the top floor of Ballast House, across the street from *The Irish Times* building. The online and print newsrooms are separate geographically as well as professionally. No opportunities are available for routine or casual mingling between the two sets of journalists through sharing space in the same building. A few online journalists contribute pieces to the newspaper—to the weekly technology page or motoring supplement or, occasionally, the features section—but do so from their desks in Ballast House.

Informality seems the defining characteristic of the *Ireland.com* atmosphere: editors, sub-editors, and journalists sit beside or across from each other. None of the editorial staff is closed off in a private office, with the physical and symbolic barriers it imposes. In this setting, conversation among colleagues comes easy; the bulk of it relates to work, but not all of it. The new soccer season has kicked off, and the website is running an online fantasy competition, with an IRL£10,000[1] prize for the winner. *Ireland.com* staff enter teams for fun. On Monday morning, after the first weekend of matches, the deputy editor and the journalist who sits opposite him chat about the performance of their fantasy teams. (In short, neither of their teams is in danger of winning, even if company policy on employees entering competitions allowed them to do so.)

Pope says, joking, "If we were any more laid-back we'd be horizontal" (interview, June 14, 2001). The proviso to the relaxed atmosphere is that work must be done on time and to the required standard.

Informality is a rather superficial difference, however. There seems a deeper motivation to the online journalists' willingness to present their newsroom's atmosphere as being different: their position within the company has a lower rank than the print journalists. This is embodied institutionally in their general employment status: lower pay, fewer benefits, fewer permanent positions. The National Union of Journalists has formed a branch for the online newsroom and is negotiating with management to improve terms and convert more contract jobs into permanent positions. A memo in the office invites online journalists to a meeting to vote on the outcomes so far. But management isn't offering parity with print journalists and, realistically, the online journalists don't expect it. Narrowing the gap is the objective, for now.

Pope reflects on the situation. "When we started the website five years ago, the terms and conditions were pretty poor. And it's still less than what our colleagues in the print newspaper earn, but hopefully that will be resolved" (interview, June 14, 2001).

He speaks without resentment, but his words confirm the underlying tension in the online newsroom: print journalists and online journalists are not equal within the organization. This has translated into a low-level attitude among the online journalists of having a healthier regard for individual print journalists (professionally and personally) than for the collective of Irish Times journalists.

Beyond material issues such as pay and contract status, another anxiety gnaws at the online newsroom and prompts the journalists to promote its virtues: the suspicion that the company's senior executives, loyal first to the newspaper, tolerate more than appreciate what the website offers and what it is trying to achieve. In discussions, formal and informal, with online journalists, a recurring theme is pride in the idea of belonging to the first generation of journalists working in a newly emerging mass medium for news, and exploring its new potentials and possibilities. In circumstances where the financial recognition of their work is lower, the suspicion of scepticism about the value of their journalism triggers self-justification.

The isolation, physical and professional, of the online operation from the print organization is crystallized in the day I spend attending editorial conferences in the online newsroom and the print newsroom. These conferences are the main channel of communication between the newsrooms as structures to gather, produce, and disseminate news.

Ireland.com holds one editorial meeting each day, at 10 in the morning. Because the journalists work in a confined space, and most of their assignments are telephone rather than field based, impromptu newsroom conferences can be arranged later in the day if necessary.

The Irish Times holds three main editorial meetings each day: in the morning at 11, the late afternoon at 3, and the evening at 6. One online journalist attends each of these meetings, to brief the newspaper on the website's news-list and to keep the online newsroom up to date with the newspaper's agenda. Occasionally, on days with big breaking stories, an online journalist will stay in the print newsroom, but as an observer more than a participant: the primary task is to keep the online newsroom informed as the newspaper moves on a major story.

First, though, is the online news conference. No print journalists are present. A brief news-list is printed and distributed.

Journalist Fiona McCann has been on duty since 6 a.m., and she explains how the news agenda has looked so far this morning. She has gone through the normal news gathering routines, such as phoning the garda (police) press office. She has surveyed news-agency copy for breaking foreign news, has moved on domestic news stories where possible so early in the morning, and has freshened the breaking-news section by removing yesterday's stories that have gone stale.

The news-editor allocates assignments. I leave with the online journalist who is designated to attend the newspaper's editorial meetings today.

As we cross the street, he sets the protocol for our attendance. "We aren't expected to say much," he tells me.

The meeting room, on the second floor of *The Irish Times* building, is functional—a large table, chairs that don't invite lounging, a television and video in the corner, switched off. The setting is ideal for deciding what needs to be done and briskly leaving to do it.

The morning editorial meeting brings together the newspaper's editor, the section editors (news, features, sport), and the online journalist to discuss their agendas and outlook for the day.

The online journalist and I are not the last to arrive, and seats are empty at the table, but he guides us to spare chairs at the back of the room, near the door.

When everyone has arrived, a news-list is handed around. With prompting from the editor, "Malachy, sports," the section editors run through their news-lists.

"Online," says the editor, after each of the section editors has spoken.

The online journalist edges his chair nearer to the table and details the *Ireland.com* news-list, a substantial portion of which is already online and could have been accessed by anyone here in the minutes before the meeting.

None of the online stories are picked up for the newspaper.

The early runner for the front-page lead is the publication of a document on Northern Ireland policing, which is being reformed as a key element in the peace process. The document's status as main story is provisional: advance copies haven't been distributed to the media, and the nature of the contents (dull or dramatic) is not yet known. The editor expects to have a clearer indication of its news value at the afternoon editorial meeting.

The editor and section editors bounce comments on stories back and forth as the news-list is dissected further. An exclusive story on Dublin Corporation (responsible for maintaining the city's infrastructure) is marked "not for breaking news." The website's immediacy, against the slower grind of the newspaper's 24-hour cycle, needs to be managed so the online journalists do not accidentally scoop the exclusives of the print journalists.

My presence, as an outsider, at a national daily newspaper's most sensitive editorial meeting of the day goes unquestioned and unremarked. I'm in open view, so it can't have gone unnoticed. To the best of my knowledge, neither the editor nor the section editors have been told in advance that I would be here.

A telling detail emerges as the meeting is about to wrap up. "Has online anything to add?" asks the editor.

Online hasn't anything to add. Online's name hasn't been used once in the course of the meeting. Others, "Malachy, sports," have. Perhaps, sitting at the back, clearly associated with Online, my presence going unremarked is not so remarkable.

Back at the *Ireland.com* office, the print news-list is passed on to the news-editor and factored into the online news-list. The breaking-news section's strength is the ability to give immediate and factual reports, and is best suited to events, announcements, and publication of reports. Opinion pieces, news-features, and analysis, which the paper will be doing, aren't considered for breaking-news. The website's editor and deputy editor also look at the print news-list, as it is the first indication of the volume and type of stories that will need to be repurposed about ten hours from now.

Throughout the day, the onus is on the online newsroom to keep in touch with the print newsroom, as evidenced by the breaking-news editorial meeting at which no one from the newspaper was present. In my three weeks of research, I do not see any print journalists visit the online newsroom.

The next two print editorial meetings are quicker, less formal affairs—progress reports on how tomorrow's newspaper is shaping up. The policing document has fallen from the front-page lead. The 6 o'clock editorial meeting, barring late breaking stories, puts the news to bed. The emphasis switches to the sub-editors, who by now outnumber the journalists in the print newsroom. The final editorial meeting gives the online journalists a clear idea of what to expect when the content appears on their system to be repurposed.

The breaking-news section will continue rolling until about 10 o'clock, when—barring a late major story—attention turns to repurposing the print content. By 6 the next morning, before people have their hands on the newspaper, the print content will be live online. The website will also have begun the production of its own breaking-news. News published on a 24-hour cycle and news published immediately sit side-by-side on the website.

Accommodating continuous and immediate publication of news required a bigger mind-set shift within The Irish Times Group than at first it might seem, especially in relation to organising journalistic practice around deadlines. Pope says, "We respond to deadlines minute by minute, hour by hour. *The Irish Times* has, basically, two deadlines per day. They have one for the country edition of the paper and another for the city edition. That's all" (interview, June 14, 2001). In a newspaper company of long standing, reporting to deadlines is a sacrosanct tradition. Working to immediate deadlines, publishing early versions of the story while the journalist continues to gather information, is a new experience within the organization.

The Webcast

When the 6 o'clock editorial meeting ends, I turn my attention to how *Ireland.com* will perform in a territory closer to broadcast than to print—webcasting. The newspaper's education correspondent, Emmet Oliver, will host a panel discussion with two education experts on the Leaving Certificate results, and will respond to emails from school students with advice on their university options.

Oliver is enthusiastic about the webcast. He sees tonight as an example of how the website can complement the newspaper: the immediacy of information this evening enhanced by a deeper analysis in print tomorrow.

Oliver, an online journalist and I take a taxi to Dublin's north inner city, to The Yard studio, where the website has been outsourcing its audio and video recording. The online journalist's role is to coordinate the webcast between the studio and the online newsroom, and to screen emails from students. Technicians at The Yard have attended to the technical aspects of streaming the discussion live on the Internet. Oliver and the panel are seated in the studio and tested for microphone levels.

Once the webcast begins, the activity within the studio is difficult to distinguish from a radio broadcast, and shows how a new media activity is often framed within older media conventions. I speak to technician Adrian Legg, who is monitoring the equipment to ensure the stream is smooth. He broadly agrees that the influence of old media can be seen in the new, but believes it is something multimedia will have to mature beyond. He says that current online video amounts to "chucking TV on the Web. It isn't really suited. We won't see it being done properly until we have video produced with web production values rather than television values" (interview, August 15, 2001).

The panellists slide comments back and forth as the webcast wins a healthy email response from students. Despite little studio experience, Oliver survives the webcast without any glitches on air. The webcast's success adds to the optimism that audio and video will enhance the website. But the step The Irish Times Group is about to take is, in many respects, farther than the step it took to set up the website in the first place. The natural territory of The Irish Times Group is the text-based medium of print. Audio and video is closer to broadcast, requires different news production skills and technologies, and doesn't offer the company the initial leverage that text gave it for online publication.

Pope is conscious of the risk of moving beyond what has proven successful, both for the website and professionally for its journalists, but is confident it will mark progress toward a more mature online news-service. He cautions, "They're [audio and video] a trial; they might not work," before adding, "There's a twofold element to them. It improves the quality and nature of the content we provide, and it also affords the journalists working here the opportunity to develop new skills…. It's very important for us to ensure the job evolves and remains challenging" (interview, August 22, 2001). Journalist Fiona McCann admits she doesn't know how the practicalities of introducing audio and video will work, but believes the online journalists must embrace such challenges (interview, August 7, 2001).

Precise details on how breaking-news will be redesigned have yet to be decided. A small recording booth will be built in the online newsroom, for routine news packages, with a high-speed line to The Yard for streaming and archiving. The online journalists will undergo voice and screen testing in the coming months. Audio reports will be introduced first, and if they prove successful, video reports will follow.

Since the Ethnographic Study

By the end of 2001, The Irish Times Group had fallen into deep financial difficulties. The Irish economy had slowed sharply, on the back of declines in the international technology sector. *The Irish Times'* circulation held, but advertising revenues diminished: rising costs pushed the Group into an operating loss.

Consolidation to protect the core newspaper business choked boom-time ambitions to expand into a multimedia company. For the year 2000, then the latest set of published accounts, the online operation had racked up a trading loss of IRL£2.7 million (over €3 million) (*The Irish Times*, April 25, 2001). The company could no longer absorb such loses while establishing *Ireland.com* as the premium brand in the online Irish news sector.

The website survived, however, but budgets were stripped and 40% of the full-time staff was fired. *Ireland.com* tightened its focus on newspaper content, breaking-news, a searchable archive, and a subscription email service. Plans to introduce an audio and video news-service were cancelled.

The strongest impact from the website's new strategy was the introduction of subscription fees to view content. For the remaining journalists, the disappointment was that fewer people were turning to their work as a source of online news. Pope said, "I think it's a shame that we've gone from having 30 to 40 million page impressions to having an awful lot less than that. But the commercial imperative is there and we have no choice" (interview, August 29, 2002).

Free content remained available, but the majority was locked behind the subscription wall. Subscription fees were set at €79 a year, €14 a month, or €7 for a week's access.

In 2007, the newspaper's financial health had improved considerably, with a rising circulation, improving advertising revenues, and a growing contract printing business. The website still charged subscription fees but continued to lose money: €180,000 in 2006, according to the latest figures available (*The Irish Times*, July 13, 2007). *Ireland.com* maintains a separate editorial team, but the Group's managing director has spoken of being in the "early stage" of integrating the newspaper editorial operation with the web operation.

In late 2006, *Ireland.com* signed a deal with Swedish company, Kamera, to provide it with video international news. The video service, *Ireland.com TV*, sits alongside the core offering of *The Irish Times*, breaking-news, and email. Despite the new video service, and some 13 years of development, the website remains primarily a text-based news-service dependant, heavily, on the content of the newspaper.

NOTES

1. Irish pounds were substituted by Euro in 2001. The amount of the prize was equivalent to 12,700 euro (in 2001, about U.S.$11,400).

CHAPTER FOUR

Print and Online Newsrooms in Argentinean Media: Autonomy and Professional Identity

Edgardo Pablo García

The case of the Argentinean newspaper Clarín *paradigmatically portrays the tensions between print and online staff in the definition of the role of the new medium and its relationship with the old. The author's emphasis on describing the origins of the online project as a key to understand the current situation gives a dynamic, evolutionary perspective that is crucial in online journalism research: everything remains under constant negotiation. In what is emerging as a common theme in research on new media news production, Garcia's ethnography captures the discomfort of online journalists as they question if their tasks are "real" journalism practices.*

EDITORS' NOTE

This chapter examines the relationship between online and print newsrooms in an Argentinean newsroom. The discussion is part of a research project examining two cases: In 2003, I spent two months in the online newsroom of *LaNacion.com* and the following year four months was spent in the newsroom of *Clarín.com*. In each case, I was present five to six days each week, and conducted in-depth interviews with almost every member of both newsrooms, observed news production routines, and accessed internal documents. Data presented here addresses the working routines in the online newsroom of *Clarín.com*, and its relationships with its print counterpart.

Methodology

As the introduction of this book observes, most of the research in online journalism based on surveys and website analyses does not inform us about how an online medium is produced, how newsrooms are organized, and the nature of the relationship between print and online newsrooms. Ethnographies, such as the ones collected in this volume, demonstrate that observation of online journalists at work is an effective methodology to address these issues. Observation also allows a different positioning for the investigator when interviewing journalists—the interviewer has more elements of judgment in order to elaborate his/her questions. As Schwartzman (1993) points out, "Because ethnographers are directed to examine both what people say and what people do, it is possible to understand the way that everyday routines constitute and reconstitute organizational and societal structures" (p. 4).

When entering the newsrooms I had, as any other ethnographer, to choose what role to play during the fieldwork. Since my investigation was openly revealed to the journalists in order to gain access to the newsrooms, my role could not be defined as complete participant or a complete observer (Hammersley & Atkinson, 1995, p. 104). I was not working as a journalist; however, I still contributed to the tasks of journalists by suggesting modifications in the editing of news agency texts. I could qualify my role as that of observer as participant or, following Spradley's (1980) categorization my role was just in the middle, a "moderate participation" (pp. 59–61), the delicate balance between out- and insider.

Qualitative research interviews are favorable when "individual perceptions of processes within a social unit—such as a workgroup, department, or whole organization—are to be studied prospectively, using a series of interviews" (King, 1994, p. 16). My formal interviews were recorded and conducted in private spaces previously agreed on with the interviewees. I interviewed 16 out of 27 journalists working there (their identities are anonymized in the quotations in this chapter), plus three executives of Clarín Global and one working at the print paper. The interview guide had topics common to all the interviewees,[1] plus a group of specific topics that approached questions relative to the position that the interviewee had in the online newsroom. In the same way, interviews with the executives of both online editions referred to specific topics of their function, and they were few topics in common with the interviewed journalists.

Context and Origins of *Clarín.com*

Clarín.com is the online edition of *Clarín*, the best-selling newspaper in Argentina. Both of them belong to Clarín Group, one of the most important multimedia corporations in Latin America. The Group also publishes a free newspaper, a sports daily and two regional papers, and has shares in the largest provider of print paper in Argentina. Clarín Group is also composed of a news agency (*DyN*), a group of successful magazines, three regional and four national television networks (including one devoted to 24/7 news), a national radio network, and an educational publishing house. Additionally, the company partially owns the two main cable providers, a film production company, and the manager of the soccer broadcasting rights. In the specific field of the Internet, the Group has an Internet Service Provider, *Prima*; the portal *Ciudad Internet*; the search engine *UBBI*; and provides broadband services in partnership with the cable providers.

Until the foundation of Clarín Global (the business unit responsible for *Clarín.com*), *Clarín Digital* (original name of *Clarín.com*) depended on AGEA SA the company operating the *Clarín* newspaper.[2] Clarín Global administers the websites and maintains coproduction contracts with Group companies that generate the content. Clarín Global is divided into two management areas, commercial and editorial, both headed by Guillermo Culell. He defined the objectives of *Clarín.com* in terms of achieving an identity which allows readers to "recognize them as different from *Clarín*." He believed that "each medium builds its identity inside the nature and characteristics of its own [media] environment."

The initial developments of any project offer the best opportunity to understand their structures and position within the institutions they belong to. The different versions of these origins by their protagonists are a very fruitful source for the researcher. In March 1996, *Clarín.com* was born under the name *Clarín Digital*. Two pages of the print edition described its main features as highlighting interactivity and providing the option of listening to broadcast news programs and football games. According to the then general editor of *Clarín*, Roberto Guareschi, *Clarín.com* did not enter into competition with the paper because "they are different media for different uses." Differentiating themselves from their predecessors was what delayed the answer of Clarín to its competitors, especially *La Nación*. According to Culell, "Toward the end of the year [1995] *La Nación* went [online]…and then we already said that we had to start with something…. We decided to take more time and we went out with *Clarín Digital* that was, already at that moment, something vanguardist, because it had audios, chat rooms, things that were not usual in an online paper."[3]

Another story on the origins offers a slightly different vision. According to Ricardo Roa, deputy general editor of *Clarín* newspaper, *Clarín Digital* arose as a strong commercial bid for a future digital market. So high were the expectations that "we had foreseen that in not many years the company's value of *Clarín Internet* was going to be higher than that of the print newspaper." Such perceptions about the commercial potential of the online market were exaggerated, but they established parameters that now condition the development of *Clarín.com*. Roa highlights that internal discussions about the project settled down in two different levels. There are, on one hand, the Group's management, and on the other, the journalists. As Roa explains, "the hype over [the Internet] as a business opportunity, gave the management a participation in this…and that made a different decision environment. This was seen as a managerial question…" This is still true today. After a stage in which *Clarín Digital* depended on the editor of the print newsroom, the creation of *Clarín Global* at the end of 2000 put *Clarín.com* under the umbrella of this new company. In the new context, Culell, the managing editor of the online venture, reported to the Internet Unit of the Group for commercial issues, but for editorial decisions he still needed to report to the editor of *Clarín*.

The relationship between the online and print newsrooms was debated during the initial developments of the website and covered two areas. First it had to be decided if the online newsroom should be autonomous of the print one, and, secondly, whether—even if depending on the print newsroom—the online should work in a physically separated newsroom. *Clarín Digital* was originally understood as a newspaper supplement. Those that postulated the autonomy of the online newsroom highlighted the necessity of developing online journalism as a specialty, a different genre of journalism. In the opposed corner where those, like Roa, who thought "*Clarín*, in its online edition, should be inside the same newsroom that prepares the print edition. On this, what we pretended was to generate a continuity of experience from the beginning, both newsrooms besides one another." When they reached the decision of making the online newsroom an autonomous one, it led to the closure of the second debate and the online staff was assigned a different location, separated from the print newsroom.

These debates establish the two big paths of the relationship between both newsrooms. On one hand, the commercial axis, originally related to the dream of big profits, and—later—to growing costs and uncertain benefits. On the other was the editorial axis, about the role for the digital edition. Both have played a part in structuring change, most importantly, in editorial policy. Even when the original decision had been to grant autonomy to the digital newsroom, that autonomy was ill-defined. Guillermo Culell told me that in 2004 the digital edition had "to a large extent functional and strategic freedom." This statement does not provide a detailed description of that freedom, but resonates with his willingness to understand print and online as two separate newsrooms, as is the case at *La Nación*.

Before moving on to the structure of the online newsroom, there are relevant aspects of the debate on autonomy that should be discussed: for what reason does the digital edition require editorial autonomy? Before answering this question, let us consider another: is this a valid question? If we discussed the digital edition of a print newspaper, we are talking about a product that comes to the world under the brand of their older cousin, the print edition. It is supposed that the editorial policies of both editions should converge, since— after all—we are talking about the same media organization. It could be argued that, under such approach, other media belonging to the multimedia conglomerate, as Clarín Group, should have the same editorial policies. However, the argument weakens itself when we consider that, at least in the case of Clarín Group, the creation of a multimedia corporation by the acquisition of diverse media did not necessarily lead to a homogenization of cultures and editorial policies; the television and radio networks owned by Clarín Group have kept their own production cultures.

But, in the case of *Clarín.com*, the online journalism genre is still in its infancy. As television—in its origins—borrowed most of its principles from the radio, online papers are still very dependent of their most immediate referent, their print counterparts. Online newspapers continue privileging text as the format for news. Potentially, online news can be multimedia. However, so far, we can only establish that their only technical potential that has been fully developed is their capacity to update information constantly 24 hours a day. This characteristic generates different production processes that provide the uniqueness of this medium. However, uniqueness does not necessarily imply editorial autonomy. Most of the online media depend on their old cousins, and as long as they are beneficiaries of the brand, their editorial policies are defined outside the online newsrooms. Later we will see how editorial control is exercised in the case of *Clarín* and its digital edition.

Breaking News and the Construction of the
Online Newsroom Structure

The breaking-news section of *Clarín.com* (labeled Último Momento, literally Last Moment) offers continuous updates of the main stories of the day starting at 8 a.m., when the first reporters arrive to the online newsroom. The origins of this service, a basic commodity among online papers in Argentina, are not clear. From the interviews, it could be deduced that in its first year, 1996, the newsroom had developed an information service for pagers called *Clarín Sport*. According to Guillermo Culell, "we were 16 hours [a day] sending messages with sport information. Then we said that, since we sent [that information] to the pagers, we [should] upload that last-minute stories to *Clarín Digital*. And we began with small boxes on one side [of the homepage], with Sport Latest News…" This service started in September 1996. In 1998, the digital edition of Clarín had breaking news in the homepage (Journalist 1).

The great qualitative leap took place between 1999 and 2000 through two major stories,[4] which established the first landmarks in the progressive advance of breaking news over the print edition contents in the homepage. In both occasions, *Clarín.com* broke from its traditional layout, which prioritized the news of the print edition, and placed Last Moment news in the central column. The third news event that definitively changed the homepage layout and the editorial policy of *Clarín.com* was, obviously, 9/11. Since then the homepage will always have two big breaking news stories up and another one below, all of them with headline and sub-headers, and then a great number of stories without sub-headers (Journalist 1).

We have already noted that in charge of *Clarín.com* is the Content Manager, Guillermo Culell, who is responsible for editorial and commercial aspects. But in terms of editorial daily decisions, the Editor in Chief, Marcos Foglia, is the journalist in charge. His authority includes all online newsroom activities, even the Publications area—responsible for the *shovelware* process of print content to the Internet. Besides Publications, the online newsroom comprises four more areas: Last Moment (LM), with three editors in two shifts, Warm Content[5] (WC), Multimedia (MM), and Community (C). One of his main tasks, shared with editors of the Country section of the website, is to attend the editorial meetings at the print newsroom, twice a day, Monday through Friday.

If we exclude those working as trainees in the Publications area, the team is made up of 31 members. Eleven of them work exclusively in the LM area.

Three reporters at LM repurpose videos from the 24/7 news broadcaster of the Clarín Group. In the WC area, four journalists work in producing the Day Report (a special multimedia report).

The Relationship between Print and Online Newsrooms

In spite of the fact that the debate on the relationship of the newsrooms was resolved in favor of the autonomy of online from its print counterpart, economic reasons and editorial discrepancies caused a different status quo, in which the online newsroom has lost a great part of its original autonomy. The relationship between both newsrooms could be studied from two perspectives. The first refers to the online edition's editorial policy, and the second to the relationships and mutual perceptions between members of both newsrooms.

The official link between both editions has implications in both strategic planning and daily routines. The former is in Culell's hands. He deals with the paper's general editor on these topics. Commercial topics are not part of this link between newsrooms because on these issues, Culell reports to the Internet Unit. From the perspective of the deputy general editor of the print edition, Ricardo Roa, discrepancies between editorial lines can be sorted out through very simple procedures: online editors' participation in editorial meetings at the print newsroom and the permanent location of a member of the online newsroom in the print one.

The Newspaper as Reference

The print newsroom carries out two daily editorial meetings. The first, at 1 p.m. and a second one at 5 p.m.[6] A member of the online newsroom takes part in both meetings. This participation is a task divided between the chief editor and the two homepage editors. In both meetings the online editors have a passive role. "We don't have participation; the [print] sections present their coverage. We could say something but these are not common situations" (Journalist 2). This attitude is in line with the objective of their presence in the meeting. They are "listeners. We are in a side and we go in order to have the panorama of news they present and [we] listen to their debate…in order to try to mimic their take on the stories [in the online edition]" (Journalist 3).

I discovered this participation of online journalists in print newsroom meetings in my first week there. One of the online editors was reading some papers. When I asked him about them, he said that those papers include the next day's print edition contents, coming from the first editorial meeting at 1 p.m.[7] Later, I discovered how important those meetings are for online journalists. On April 28, Marcos Foglia returned from the first one and immediately started to distribute the paper's agenda among journalists of Last Moment. Then, he explained which the two hot news topics for the paper were. Online journalists were supposed to pay attention to these topics when receiving information through the news agency wire service. But this is not the only way the print newsroom exercises its influence/control over the online one. Some days later, on May 24, coming back from the 1 p.m. meeting, Foglia not only distributed the handouts with the agenda but also instructed the editors of the Country section "to kill" one of the prominent news stories on the homepage. Asked why, Foglia replied that one of the most influential journalists in the paper suggested to him that story was not so important.

They attend the second meeting, which concludes at 5 p.m., because this one generates changes in the homepage. Those modifications are carried out in order to align the editorial perspective of the website to the one defined by the print newsroom in the second meeting. For instance, Foglia highlights that "an economic topic that arose at 11 o'clock in the morning, at that hour there was a specific vision [on it], but at 4 o'clock in the afternoon when the newspaper specialist arrived with another vision, and then the focus was modified in relation to the 1 p.m. meeting…. We, at 6 p.m., adjust [the approach on the website] if we are very far from the editorial approach that prevails in the newspaper."

It is 7:10 p.m. on May 14. One of the online editors has just arrived from the 5 p.m. meeting at *Clarín*. He starts talking with the journalist of the night shift, who has just started to work. He asks him to move the news story on pensions from the bottom of the homepage to the upper right corner, the most important area. He tells me the change is based in a print newsroom decision to include that story as one of the most relevant ones in tomorrow's front page. For those who do not participate in the editorial meetings, it is clear that decisions made there govern their work at the online edition. This is particularly visible when the online editor returns from them: "When they return from the meeting they say 'hey, make this topic stronger [on the homepage] because they are to going to give it attention, and kill that one because, in the end, nothing happened'. Like it or not, although there is not a direct command…we follow them [the print newsroom] a lot" (Journalist 4).

This pursuit of *Clarín*'s editorial policies is also clear during the day. In a more informal way, consultation with print editors is a permanent practice. However, this practice has its weak points. The most obvious is that print editors usually arrive to the print newsroom several hours after the online journalists do. Roa refers to this gap when he notes, "we are not [here] in the morning, and then there is a [online] newspaper they publish from the early morning until we begin and another newspaper in the afternoon. A newspaper that walks with the online compass in the morning, and a newspaper that walks with our compass in the afternoon, or relatively with our compass, with some problems every now and then."

The print newsroom's compass gains more importance when news stories are linked to economic and political issues. The consultations strive to set the news stories' focus, tone, and relevance. The idea is "seeing how the newspaper approaches it in order to avoid giving a different focus...," because when those types of stories appear "we have to think how they will be seen by the newspaper. The central issue is to see how much importance the paper gives to the news story" (Journalist 1).

It is April 26. A journalist at the Country section is looking at wire stories. He discovers a statement from a former civil servant about a former secretary of state under judicial investigation. He considers this an important story. He cuts the wire text and pastes it in a Microsoft Word document in order to work on it. Then he uses *Clarín*'s search engine in order to find previous stories about the issue in the print edition. When I asked about the reasons for this process, he replies that it is necessary to add some information and to perceive "the tone" of the newspaper's prior coverage. Knowing the tone, he can add "the *Clarín* spirit" to his story. He also points out that despite being "two different products, both belong to *Clarín*."

Not only is politically relevant news confirmed with the print newsroom, but also those relevant to the Group's economic interests. Especially when there are "news that touch companies that perhaps are sponsoring the paper, [those] are things that have to be checked up" (Journalist 5). That type of "sensitive" news is generally discussed by journalists at the print editorial desk. In the case of minor topics, journalists check with print journalists involved in those topics.

On June 4, Argentina President Nestor Kirchner denounces several economists and politicians of "abuse of public services" during his visit to a Patagonian city. One of the TV channels is broadcasting the presidential speech but none of the online journalists watch it completely. They are not sure about the entire content. They start a small text, an "alert," with one of Kirchner's statements. Immediately, they receive a news agency story that confirms Kirchner's accusations. However, editors are not satisfied because the wire story came from the state news agency. Three editors and one journalist start a debate about what to publish. They have doubts because *DyN*, the agency owned by Clarín Group, did not send a wire story. They decide to check with their liaison in the paper's newsroom, but he is having lunch. Then, they make contact with one journalist in the print newsroom. He tells them to keep quiet, assuring that the alert is good enough. Some of them are not sure about the advice, but Foglia says they have to follow the paper's editorial policy. Minutes later, *TN*, the 24 hours channel of Clarín Group broadcasts an interview with the President, who repeats his previous statement. Now it is confirmed and they finish the story, with the names and the source—the state news agency—because the main online competitor, *Infobae*, has already published the story.

We might conclude from this episode that 1) print newsroom editorial positions have the upperhand, especially in political and economic stories, which are not only delicate ones but also essential in the paper's agenda-setter role; 2) news agency stories are extremely important to produce stories and, without them, options are to check with TV news channels or with print journalists. Consultation with print counterparts is not only about specific facts (e.g., the President's statement) but also about the editorial take on the issue.

Bonds and communications between newsrooms are not a one-way street. Print journalists call to demand changes in online stories.[8] However, the interest of print journalists in the editorial policies of LM is not homogeneous. On the contrary, it only refers to news stories in the Country section (which includes politics and economy news). According to one of my interviewees "in The World and in Society [sections] they don't enter, they are very far away …perhaps some stories on the homepage based on a wire story are wrong, and they know but don't call to warn us" (Journalist 3).

Having a permanent representative of *Clarín.com* in the print newsroom began in 2002.[9] There were several reasons for this. First, to develop economic stories for LM, since given the complexity of those stories, it was convenient that one online journalist was in contact with print journalists specializing in that area. Second, it was also the case of having a nexus "between the dotcom and the paper and…someone told me too, it is a question of having a presence, a public relations work inside the print newsroom, …[so they can recognize] that the dotcom exists…" (Journalist 5). Finally, but not less important, it is the question of "being able to check up information that because…we mainly work with news agency stories, and we didn't have possibilities to check them up for complicated topics, …to call [to the source] in order to decide if it should be published or not" (Journalist 5).

The presence of this online journalist in the print newsroom achieved more importance since the online newsroom reorganization in 2004. The definition of the Prime Time, from 10 a.m. to 6 p.m., concentrates human resources in that period, leaving the morning (8–10 a.m.) and night (6–12 p.m.) shifts with few journalists. A few months after the reorganization, Culell discovered this imbalance: "we realized…that the night was very important because it was necessary to align ourselves with the deadline of *Clarín*, …for a reason of convenience of publication and also for a question of…I would tell you of information consistency."[10] A LM story published at 1 p.m., which went to a secondary place in the homepage during the day, may later return to the first places, according to print newsroom editorial decisions. In Culell's vision, this requires an "editor [the online journalist based in the print newsroom]…one that could touch the homepage, enter in that."

Mutual Perceptions

The mutual perceptions between both groups of journalists offer a second level of analysis of their relationships. It was impossible to receive consent from the print newsroom in order to carry out interviews with their members and observe them in their daily tasks. However, the deputy editor, Ricardo Roa, provides his point of view as a manager of the print newsroom. Roa highlights that the main skill of an online journalist refers to the presentation of content, to add video, audio, etc., and this "is not a skill strictly related with content production." Referring to information presentation, the online journalist would become an "administrator of information produced by another one. That is not a [journalistic] skill. The ability of a journalist is to know how to give value to the information." And according to his argument, it is obvious that a journalist of the online edition, associated with the permanent updating of breaking news, is handicapped in producing his own information since he lacks sources. In that, the digital journalist differs from the print one and therefore "he is not prepared to differentiate news from what it is not news." The fact that online journalists lack sources is not related to lack of authority or due to institutional discouragement. Working routines do not allow them to make that kind of contact. They do not have time to pursue sources and, for those with previous print experience, this is completely disappointing. Although these judgments would seem to move online journalists away from the category of journalists, Roa thinks that it is not the case. According to him, they are still journalists.

Roa's views could be divorced from perceptions of his colleagues at the print newsroom, but not so much from the beliefs of journalists at *Clarín.com* on how they are perceived by their counterparts. Although they believe that the original perception is changing, many recognize that original stigmas still survive. Some consider that "we in the paper were half *kelpers*[11] and it is still the case, though is changing" (Journalist 1). This journalist attributes the perception to the fact that their print colleagues believe that work in the online newsroom is of minor quality because "it is different to their work" (Journalist 1), or as pointed out by another journalist, "they see that [news] are not produced here as produced in print journalism, and there they are right" (Journalist 3).

That stigma is associated with the specific reality of a newsroom that arose in a "nonprofessional [way], improvised" (Journalist 3). It is that concept, pasting wire texts instead of talking with sources, that determines, to print journalists, the minor category of online ones. As it was stated by an online journalist: "they know which is the mechanism, that we do not work with sources, that we work with wire stories and…I, unfortunately have to say, we are 'clipping' wire texts" (Journalist 5). And that characterization of the tasks of online journalists distances them from the proper journalistic work because "we didn't generate content, you cannot have an exclusive and that also, I think, is the salt of all this, of journalism" (Journalist 5). Under these conditions, it is not strange that some conclude: "I have the perception that *Clarín.com* is not *Clarín*, we are like a minor brother, half stupid, because…I feel that from the [print] newsroom, from the newspaper, there is a vision toward us, as a medium, that devaluates it, our work. In fact they do not appreciate online journalism in itself…they consider it as a sub-genre" (Journalist 4).

Other members of *Clarín.com* suggest that print journalists consider them as a "second category", but that should not be so strange, since "all print journalists think about the rest of journalists as second category [journalists], and that is because of the social role of newspapers, setting the agenda" (Journalist 6).

In the particular case of *Clarín*, what originally could have been scorn is also competition. According to some members of *Clarín.com*, the restless increase of LM news services is now a threat for the print edition and print journalists fear the online edition could take information from them or leave them without a focus. But, for most of them—as we highlighted at the beginning—the relationship is improving from new instances of collaboration, like the multimedia special news package carried out for the War in Iraq, or the requests from the print newsroom for online opinion polls that could be used by print journalists in the paper edition (Journalist 7).

Conclusions

As an information sector in its infancy, online newspapers have been exploring different models to deal with convergence and organizational issues. From multimedia newsrooms producing content to different media outlets inside the same media group, to strictly separated newsrooms. In this chapter, I wanted to offer the main features of the relationship between print and online newsrooms at the heart of the online production process. As we have seen, this relationship has been changing through the years at Clarín Group, due to an ill-defined autonomy for the digital edition. Commercial and editorial issues have come together to create a complex bond between the two institutions.

After some time, a new model was defined in order to clip the original autonomy of the digital edition. Institutional demands to align editorial policies have built online production routines around the print newsroom editorial meetings, where online journalists play a passive role, learning the paper's agenda in order to reproduce it in its own edition. Besides these meetings, the online newsroom receives clear suggestions to cut or "deflate" stories, which were prominent in their editorial judgment. But online journalists are not only waiting to receive instructions from their counterparts. They know that "sensitive" news stories, those that can affect the paper's standing in the public arena, are obvious candidates for consultation with print journalists. That is why the alignment with the print edition, through institutional channels or informal ones, is especially manifest around the Country section—political and economic news.

The digital newsroom's subordinate position is also clear in the self-deprecation of online journalists. In the last part of this chapter we have seen how they see themselves as "*kelpers*," "half stupid," and "minor brothers." They know perfectly well that their daily job is far away from the routine task of talking with sources and searching for information. They know they are not producing information as those in the print newsroom do. Some of them despise their print cousins' perceptions and consider them afraid of the online edition becoming a competitor of the print edition. The general feeling is disappointment.

The relationship between newsrooms has both levels: editorial policy and cultural aspects. The first one has been developed into specific routines. The second is still a quiet conflict that will be difficult to resolve.

NOTES

1. The topics discussed with online journalists in my interviews included their professional education, the history of the online news website they worked for, their opinions about interactivity, hypertextuality and multimedia features of the website, their considerations about journalism as a profession, their professional self-perception, the characteristics of the audience, the strengths and weakness of the website, etc.

2. At the end of 2000 the management of Clarín Group faced a process of international offer of shares. In that moment, when the dotcom bubble was still important, the management perception was that making the digital endeavors of the Group into one structure independent of AGEA SA, could contribute to grant more value to Clarín Group. The second reason behind the creation of Clarín Global was to strengthen synergy between those digital endeavors belonging 100 percent to the Group.

3. But the experiences of Clarín with digital technologies had already begun in 1995. In August that year the newspaper celebrated its first 50 years, and they decided to publish a book and a CD-ROM. These two materials, thought originally for a noncommercial distribution, evolved and the CD-ROM was marketed with certain success. This modest commercial success led to the creation of an area of digital media, where Guillermo Culell and Julio Gallo worked, in order to produce CD-ROMs. They only developed one other CD-ROM and, according to Culell "Then there came the Internet. We did a first experience with the University of Buenos Aires, for the 1995 elections…. We did a very primitive site with people of [the Faculty of] Sciences…. They put the technology and we put the journalistic content…with information on the elections. We had an enormous amount of visits for that time in which almost nobody knew about Internet. Starting from there we began to think."

4. In September 1999 a commercial airplane crashed during take off from Buenos Aires airport, killing 67 passengers and crew. In October 2000, the Nation Vice-President renounced to his position after having demanded an investigation on Senate's corruption.

5. The Warm Content area includes the Connections and Television sections of the website, as well as the multimedia Day Report, all located in the right column of the homepage.

6. The first includes an evaluation of the morning print edition and the current day coverage is discussed. The second meeting allows to evaluate the development of news stories and to sketch the first draft of the front page for the next day edition.

7. Usually, online editors of the Country section receive a draft version of the paper's political and economic agenda before 1 p.m. The online journalist at the print newsroom sends it by email. This draft is later confirmed or modified during the meeting.

8. This is not an uncommon feature. As Marcos Foglia pointed out to me, if one member of the editorial desk at the print newsroom tells him that some news story is "inflated," the online journalists do not usually oppose that demand and proceed to make adjustments as directed.

9. This permanent representative is spatially located in the the Country section at the print newsroom, from where he has direct access to the Politics and Economy editors, and to the members of the editorial desk. Besides, this journalist also produces videos with print journalists.

10. This brought a change in the working hours of that journalist assigned to the print newsroom, from 12 to 8 p.m., to 4 to 12 p.m. He was also appointed as Editor of the Country section, so he can carries out the last-minute changes in the homepage.

11. "Kelper" refers to inhabitants of Falkland Islands. It is used contemptuously in Argentinean slang.

CHAPTER FIVE

News Tuning and Content Management: An Observation Study of Old and New Routines in German Online Newsrooms

Thorsten Quandt

Online journalists devote most of their time to selecting and editing news agency wire stories; they do little original reporting and engage in few technical tasks. This is the profile offered by this summary of a large study of working routines in five German online newsrooms. Ethnographic data gathering was uncommonly systematized and processed into quantitative and network analyses. Newsroom organizational structure, size and location offer valuable context to interpret the results. Especially noteworthy, beyond Quandt's innovative research methodology, are his descriptions of the relatively small size of the online teams and the centrality of content management software in their news production routines.

EDITORS' NOTE

Online Journalism Research in Germany

In the second half of the 1990s, online journalism was one of the hottest topics in journalism research and the media industry alike. Many visions of revolutionary new forms of multimedia and online journalism in a converging news environment were discussed at that time. As Joshua Quittner described it in a much quoted article in *HotWired*, there were expectations of a "whole new journalism" (Quittner, 1995) that would change the way news is produced and perceived. Some years later, though, based on his day-by-day experiences with online journalism, Quittner noted that his early prognosis of a revolution in journalism was wrong.[1] A lot of the overly enthusiastic prognoses might be based on a logic *short circuit*: technological options are often directly extrapolated into real life social use—without taking into account that media as social institutions exist in complex contexts (Rühl, 1998; Görke, 2000). Not everything that can be (technologically) done will be realized (in the realm of the social) or will be done with success (see chapter 1). Furthermore, many visions of a "new" (online) journalism have no empirical basis whatsoever; and research findings from the journalistic context of one country cannot be easily transferred to other cultural and national contexts. What is true for American online journalism, for example, need not be true for German online journalism—which is discussed in this chapter.

There are a considerable number of academic works on online journalism in Germany (Kopper, Kolthoff, & Czepek, 2000; Neuberger, 2003; Quandt, 2005). A lot of these are theoretical and speculative, especially the early works from the mid- and late 1990s which were focusing on the innovative aspects and the technological options. Online journalism was often discussed in a technological context, with the online news worker being less of a journalist but more of a "creative computer freak" (Benker, 2001, p. 61), thus, a technology centered job.

A few smaller newsroom studies (Schmitt, 1998; Wilke & Joho, 2000) sketched a different image of online journalism. While they observed some technological aspects in online journalism, these tasks were mostly performed by technical specialists in a newsroom context, while the journalists were focusing on the traditional news work. However, these studies could only give hints of what online journalists in general do, since they were limited to one specific newsroom and a short observation time. Much broader information on online journalists can be obtained from several larger survey or interview studies (Kuhnke, 1998; Mehlen, 1999; Neuberger & Tonnemacher, 1999; Neuberger, 2002; Meyer, 2005; Quandt et al., 2006). These studies offer data on the number of online journalists, their attitudes, viewpoints and motivations, the organizational structures of the respective media companies, and some self-estimations on the news work. The latter have to be interpreted with caution, though: It has been shown in the past that interview studies and real-life observations in journalism lead to differing descriptions of work reality (Altmeppen, Donges, & Engels, 1999, p. 158). The answers in interview studies are based on rational reflection of past actions; so the respondents are very often forced to construct a memory representation of processes that were partially or fully below their consciousness or control. Neuberger realizes this general flaw in interview studies with online journalists, especially when it comes to the daily work routines, and comments: "Not much can be said about the work context in online journalism" (Neuberger, 2000, p. 37).

In order to fill this void, I conducted an ethnographic study of five German online newsrooms. Its design will be discussed in the next section, followed by a description of the work contexts in the five newsrooms, some selected findings from the observation of the daily work routines of online journalists in these newsrooms and a discussion of possible future research paths.

The Ethnography: Research Interest, Theoretical Approach, and Study Design

Based on the lack of research on the actual news work in online newsrooms in Germany, an explorative approach is necessary to get first impressions of the field. Accordingly, the study is aimed at the description and analysis of the emerging journalistic work, so a theoretical approach based on action theory seems to fit the given problem in a most natural way.

Such an approach was developed for this study, inspired mainly by the works of Schütz (1981, 2002) and Giddens (1984). It portrays how work patterns are produced and reproduced in daily work routines, while at the same time these emerging patterns form and organize further work (see Quandt, 2003, 2005, 2007). To precisely analyze this pattern building process, each individual action is analytically divided into its constituent elements first: the action core (the most basic, largely undefined type of action, e.g., "writing," "reading," etc.), the resources used for the action, the persons involved in the action (like contact persons, colleagues, etc.), time and space conditions of the action as well as the reference frame of the action (for example, an action can be framed as happening in a private or work context). These constituent elements can form synchronous relational patterns and typical configurations of actions (i.e., a resource is very often or always connected to another person or a specific context in a given type of action; such connections are called *associations*), as well as time-based patterns (one element always follows a specific other element; these patterns are called *sequences*). By working, the journalists reconstruct the idea of their work by reinforcing some patterns, while others are not confirmed through repetition. Furthermore, through this everyday practice, the journalists will also build up comparable webs of sense-making relations in their "stock of knowledge" (i.e., connections in memory that allow for a complex behaviour (Schütz, 2002, p. 153)—therefore, the individuals are actually sharing some *meaning*, at least to a certain extent, which might lead to a common "orientation horizon" for their work.[2]

From this theoretical basis, an observation study of six individuals in five online newsrooms (*Netzeitung* in Berlin, *FAZ.net* in Frankfurt, *SVZonline* in Schwerin, *tagesschau.de* in Hamburg, and the Capital office of *Spiegel Online* in Berlin) was conceived and carried out during 10 weeks in the summer and fall of 2001.[3] These newsrooms were selected for various reasons; basically, they represented different approaches and backgrounds (parent companies from different media backgrounds, varying age of staffers and size of the newsrooms); these are explained in more detail below. However, all the observations took place in the political departments (if existing); this was done to prevent biased findings in this respect (for example, it would not be sensible to observe and compare journalists from the sports department of medium A with the political journalists of medium B).

Derived from the sketched theoretical outline, a codebook was developed for the empirical study, consisting of roughly 250 codes in several categories to describe the constituent elements of the journalists' actions. By combining these elemental codes, each possible variation of a journalist's action could be coded—basically, the codes and their relations formed a description language for human action. In order to realize the coding in real time, all the codes had to be memorized beforehand, and coders had to try out the actual coding process in pretests. In addition to the closed instrument, open fields on the code sheets allowed for an additional description of the actions.[4]

In addition to this, alternative ways of information gathering were used. To provide a better understanding of the work processes, observational diaries were set up to write down open questions that could be answered during interviews with the journalists and their chief editors. Last but not least, photographs of the workplaces were taken and floor plans of the work places were drawn in order to get an impression of the work contexts.

Using these instruments, a detailed description of the work processes was possible. The net observation time without free days (holidays, weekends, etc.) was 44 work days, corresponding to more than 400 hours of observed activities. During that time, 11,671 actions were described using the code language, of which 10,826 were further analyzed (private activities were excluded from the analysis). In the resulting data matrix, each act consisted of about 50 variables that would describe its constituent elements in more detail. Therefore, the data basis for further analysis was very large, and the subsequent data restructuring and analysis (using statistical tools, data mining procedures as well as qualitative interpretation) took several months.

The Online Newsrooms: Space, Time, and Organizational Context of Work

In this section, the organizational, spatial, and time-related contexts of the work in the observed newsrooms will be described; since the observations took place in five newsrooms, not all of the details can be explained here (for more in-depth discussion, see Quandt, 2005). Rather, I will focus on an overview and some general trends that apply to most newsrooms.

- *Netzeitung* is the only independent online newspaper in Germany, not an affiliate of a larger traditional media company. It started in 2000, with the structure of a print newspaper in mind, with several beats, and it had 30 journalists at the time of observation.

- *tagesschau.de* is, since 1996, the online counterpart of a news program of the public TV network ARD. The online newsroom is a more or less independent structure within the television building, with its own chief editor and 15 journalists.

- *FAZ.net* is part of the multimedia strategy of the *Frankfurter Allgemeine Zeitung*, one of the biggest German quality newspapers. Launched relatively late, in 2001, it has a journalistic staff of 30.

- *Spiegel Online* is the Internet offspring of the most important political weekly in Germany, the *Spiegel*. In 1994 it started as one of the first news websites in the country and is the current market leader. The online newsroom of 30 journalists[5] has its own editor-in-chief, with significant autonomy from the magazine, as they produce breaking news while the magazine concentrates on in-depth reporting and analysis. *Spiegel Online* also has an independent political department in Berlin (the observation took place here), where five online journalists are working in close cooperation with the main newsroom in Hamburg.

- *SVZonline* is the only of the observed cases that has not formed a fully developed online newsroom. In 1995, it was the first newspaper website in Germany and still today it is limited to shoveling the content of the *Schweriner Volkszeitung* (a regional newspaper in the Northeast) to the website, with a staff of two.

It is obvious from the above description of the online newsrooms as (sub)structures of bigger companies that they will also differ in their "internal" structure and the work flow organization. As a rule of thumb, the larger the structure, the more diverse the internal specialization, resulting in sub-units (like work groups preparing specific beats, etc.) and sometimes formalized sections or even separate dependencies (like the political department of *Spiegel Online* in Berlin). The largest three, *Netzeitung, FAZ.net,* and *Spiegel Online,* developed such sub-units. *tagesschau.de* does not have an internal subdivision, for several reasons: it is primarily focusing on political news (much like the TV counterpart), the staff numbers are still smaller than in the aforementioned newsrooms, and the chief editor believes that journalists should be "generalists." That said, there are journalists who are responsible for special topics and are usually referred to when such topics have to be covered. It is worth noting that in the four larger units under analysis, the specialization also coincides with a hierarchical structure of sorts: all of these online news sites have a chief editor and a so-called "CvD" (Chef vom Dienst, who is basically the coordinating editor in charge of a work shift; this job is usually rotating between the most experienced members of the newsroom). In two of the analyzed media, *FAZ.net* and *Spiegel Online,* there are also section heads for the political departments. Nevertheless, in all the observed media, even the largest ones, the journalists call the hierarchies "flat"; in the interviews, they usually compare them to much steeper hierarchical structures in traditional news media like newspapers and TV stations.

The size of both the editorial staff and the content it produces also results in a specific design of the production workflow. The bigger media all rely on shift work: there is at least a day shift with all departments preparing news, and a night shift with a reduced number of editors (sometimes only one) checking the output of the news agencies for relevant news.[6] Not all of the observed media have a full 24-hour shift system, though—for example, at the *Netzeitung,* they have a "production hole" of approximately five hours between 1 a.m. and 6 a.m., due to minimal readership during that time. Shift work and specialized work groups/departments also lead to the need for coordination, which is realized in all of the four bigger news media through formal editorial meetings with the chief editor (or its deputy), other informal, mostly short meetings, sometimes several times a day, and the role of the coordinating editor.

However, everything that has been said so far only applies to the bigger, "full" newsrooms. *SVZonline,* due to its minimal journalistic staff, has no internal division. Furthermore, working in shifts is not possible or necessary. Basically, there is one online journalist who is integrated into the hierarchical structure of the newspaper. He reports directly to the chief editor of the newspaper. The other member of the online team is devoted to marketing tasks.

It is interesting to note, though, that all of the observed newsrooms lack one feature that has been hailed as one of the main assets of media production in a digital environment: cross-media or convergent production. Some media had TV or radio counterparts (*tagesschau.de*, *FAZ.net*, *Spiegel Online*) at the time of the observation, and most of them (with the exception of *Netzeitung* and *tagesschau.de*) have a print correlate. Still, the cooperation is limited to the integration of *shovelware* in most cases (which is edited and integrated by interns or by the journalists themselves). So clearly, the online newsrooms are mostly independent from the other media channel's newsrooms, and not integrated in a convergent multimedia newsroom.

Newsrooms are both social and physical spaces. The organization of work and relations between the subjects in work contexts is dependent on the available space resources, and to a certain extent, these space resources are organized according to the needs of work organization—so the relation between spatial organization and work structures is recursive. As most online newsrooms are "new," they were planned on the basis of the expected needs, and sometimes altered according to the real-life needs identified afterwards.[7] Obviously based on similar considerations, the observed bigger newsrooms look very similar and have a comparable structure (see fig. 1 for an example). People working for the same beat are usually sitting together at larger tables (three or four persons), so that they can easily coordinate their work. All of these tables are located near the central news desk, where incoming agency news is analyzed, filtered, and distributed to the various tables/journalists. So this is the space where the daily editorial routine takes place.

Furthermore, the bigger newsrooms also include a special server room (not visible in fig. 1) and work places for web designers and/or the technology staff (upper right in fig. 1). Meeting rooms are separate from the central newsroom, which is usually too loud for more than informal meetings. Additionally, the chief editors (and sometimes their deputies) have an office that is separate from the newsroom; in some cases, also the secretaries or editorial assistants and the marketing staff have separate offices.

It has to be noted, though, that newsrooms are not only spatial resources, but they also offer and spatially organize the (technical) equipment for the journalistic work. So it is not surprising that similar room structures also correlate with comparable technical equipment. Naturally, the most important resource is the computer: Journalistic work in newsrooms usually takes place in front of a flat screen. In the four bigger newsrooms, the journalists use various content management systems (CMS) that are the central production tools, offering (sometimes limited) word processing functionality, an access to archived information in a data base (texts, and in some cases also pictures), layout and publishing functionality. Usually the latter is limited in the journalist's CMS user profiles, and only the central news desk or the coordinating editor have the full publishing rights, so they can control which articles are published where, and how the website looks. In addition to the CMS, the journalists use standard word processors, mostly because the CMS do not feature full word processing capabilities (like spell checking, saving of various versions, etc.). Other central tools for the online journalists are the digital news agency services. As a standard in the observed bigger newsrooms, they are available as client applications at the journalists' work places. The journalists can monitor the incoming news agency material in these applications, filter them according to their needs (for example, by using keywords, they can find all news items regarding one specific topic) and export the material to the CMS or the word processor for further editing.

By now, it should not come as a surprise that the *SVZonline* journalist differs in all of the above aspects: This journalist does not use a CMS and an agency service. The former is too expensive in relation to its use for the small website, and the latter is not needed because the website largely consists of *shovelware* from the print newsroom. Instead of an industry standard CMS, the journalist uses special hand-coded tools to semiautomatically publish online content. Furthermore, he edits material with online editors, word processors, and even Quark XPress (due to the format of the material produced by the newspaper journalists). This already hints at a differing work profile—which will be described in more detail in the following sections.

The Journalistic Work: Actions, Patterns, and Networks

In the following sections, an overview of the central observation results are given, both in statistical form as well as in the form of network analysis outputs that reveal the relations between the action elements. It must be noted, though, that this is not meant to be reductionist or to downplay the many complex

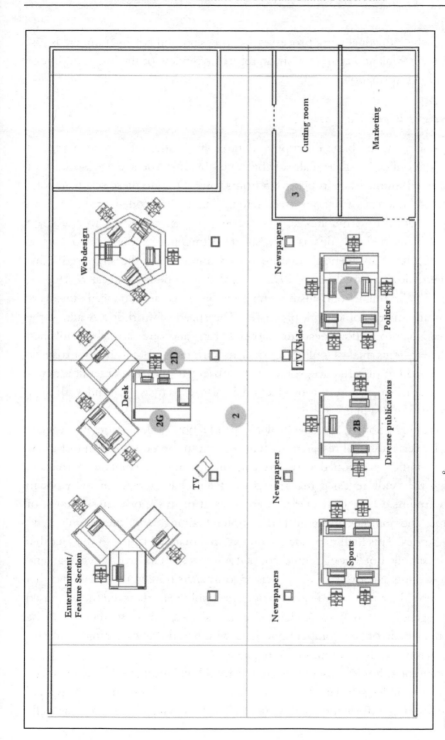

Fig. 1: A typical online newsroom: The Netzeitung [8]

details of the individual work processes—expressing findings by numbers is just the most convenient way to give an aggregated overview of the huge amount of observation information.

Analysis of Work Categories

When looking at the basic descriptive data of the study, the characteristics of the work in online journalism already become obvious: The average workday of the observed journalists is roughly 8.5 hours long. Despite these relatively long days, the frequency of actions is extremely high. One coded action had an average duration of 2 minutes and 14 seconds. When compared to studies in other journalistic fields, this is very short. Altmeppen, Donges, and Engels (1999) calculated an average duration of 4.8 minutes for the supposedly fast radio journalism (Altmeppen, 1999, p. 115; Altmeppen, Donges, & Engels, 1999, p. 72). So online journalism seems to be very fast and "minced" into very small, nearly microscopic work processes. The speed of work is a result of the disappearing production deadline: articles can and are actually published immediately—this makes online journalism different from other journalistic areas. As a point of reference, some of the observed journalists mention news-agency journalism, where the production is also extremely fast and without a real production deadline.

Interesting enough, the speed of the production cycles—from the agency material, over additional research, writing, editing to the publishing in just a few minutes—is not mentioned very often in the literature on online journalism, although the study revealed that this is the main characteristic of the work in the newsrooms. Other characteristics are a result of this. For example, most of the time, the journalists are just regrouping, editing, and fine-tuning news agency stories. This high dependency on external material is due to the fact that there is simply not enough time for a lot of research, cross-checking, and original writing. Nearly all of the observed journalists note in the interviews that if they took their time, the readers would probably read the same news (based on raw agency material) somewhere else in the meantime—so they try to be faster than their online competition. This wire-based "news tuning" will also become visible in the distribution of the action categories.

A first look, however, seems to hint at traditional journalism. The overview of actions includes standard journalistic activities, like communication, searching and selecting information and writing (table 1). What's obvious here as well:

Table 1: Action categories' share of the activities' cumulated time per person (in percent)

	NZ A	NZ B	FAZ	SVZ	TS	SPON	Average per journalist
Searching/Selecting	36.4	39.8	28.6	22.7	35.6	29.2	32.1
Text production	28.5	26.7	30.5	6.9	17.9	18.5	21.5
Production tasks	2.7	1.4	1.0	28.7	1.9	1.4	6.2
Technology-related tasks (no production goal, i.e., repairing)	0.7	0.4	0.4	0.3	0.3	0.1	0.4
Communication/face-to-face	12.6	13.7	18.0	19.3	29.5	20.2	18.9
Communication via media	14.5	12.2	17.9	14.8	5.1	24.7	14.9
Organizing	0.9	1.0	0.7	2.8	0.1	1.4	1.2
Other	1.8	1.7	1.3	3.0	8.0	3.5	3.2
Moving/walking around	1.9	3.1	1.4	1.6	1.7	0.9	1.8
Sum	100	100	99.8	100.1	100.1	99.9	100.2

Time basis: *Netzeitung* journalists: NZ A: 36:39:25 h; NZ B: 51:04:00 h; *FAZ.net:* 79:57:45 h; *SVZonline:* 95:42:40 h; *tagesschau* (TS): 77:47:55 h; *Spiegel Online* (SPON): 64:44:05 h; Overall: 405:55 h

Table 2: Text production: share of the activities' cumulated time per person (in percent)

	NZ A	NZ B	FAZ	SVZ	TS	SPON	Average per journalist
Original text production	9.7	6.5	6.4	0.3	2.6	5.7	5.2
Editing texts (own text, not published)	4.1	6.6	11.3	0.1	1.3	7.3	5.1
Editing texts (already-published text)	5.8	3.2	0.5	1.6	9.3	1.4	3.6
Editing texts (*shovelware*, parent company)	-	-	-	1.9	-	0.4	0.4
Editing third-party texts (agency material)	2.8	3.9	6.9	0.3	2.1	-	2.7
Editing user input/letters for publication	-	0.1	-	-	-	-	0.0
Notes	6.0	6.4	5.3	2.7	2.6	3.5	4.4
Other	-	-	-	-	-	0.2	0.0
Sum	28.4	26.7	30.4	6.9	17.9	18.5	21.5

Time basis: NZ A: 36:39:25 h; NZ B: 51:04:00 h; FAZ: 79:57:45 h; SVZ: 95:42:40 h; TS: 77:47:55 h; SPON: 64:44:05 h; Overall: 405:55 h

Multimedia production tasks are not of any importance to the journalists, so this clearly contradicts some early predictions of the journalist becoming a "multimedia news producer" in the online environment. The only person who has to cope with a considerable amount of production tasks is the SVZ editor—as described above, he has to transfer the *shovelware* from the print newspaper to the online version, and he has to manage the website without a dedicated content management system. So this is certainly not the "new" journalism from the early visions of a revolutionary online publishing, but the effect of an economic decision to keep the online investment at a sensible level, fitting the regional reach of the newspaper.

As said above, the overall distribution of the main categories indicate a rather traditional work pattern. This fits the self-description of the online journalists, as the *Netzeitung* journalist B notes in the interview: "In no way do we perceive ourselves as being technicians or programmers." However, a closer look reveals that there are differences to other forms of journalism. For example, there are several forms of writing, and it is important to differentiate between them (table 2).

The detailed analysis indicates the usual amount of time devoted to writing notes (one fifth of the writing time), and about the half of the text production time is devoted to original writing and editing of one's own texts. That said, there are some differences between the journalists: While the journalist NZ A writes a lot of articles (due to his preparing a special background package including various text and text boxes during the observation time), other journalists devote less time to the writing. While NZ B, FAZ, and SPON can also be considered "writing" journalists, the *tagesschau.de* journalist is editing much more, also already-published material—this is due to her being the coordinating editor for several days during the observation time, thus having a lot of coordination duties. Furthermore, the SVZ journalist writes nearly nothing, which can be explained by the work profile already described above.

What's interesting here, except these individual specifics, is the high level of editing visible during the observation. The journalists edit, cut and rewrite their own texts, agency material and already published material—something that is certainly not the case for other media, where already published material is "out of the door" and cannot be changed anymore. When asked about the high level of editing in the interviews, some of the journalists indicate that they are aware that online journalism is highly agency dependent and, to a certain extent, "secondhand" journalism.[9]

This finding is also supported by the detailed analysis of searching and selection tasks (table 3): Here, the focal points are on reading print material, looking through one's own website, searching/selecting agency material (offered by the computer client of the news wire service), and surfing the Net. "Reading" print material typically indicates print outs of news agency material or websites, so again the agency dependency is visible here, as well as the use of online material as information sources. It is interesting to note that online material is mainly coming from competing news sites; the journalists constantly monitor the output of their (German) competitors in order to check their news in relation to what the others offer. This corresponds to the high level of watching one's own website—here, the journalists a) compare the articles of other media with their own, or b) check their own articles for mistakes or missing information after the publication. If they find problems, then they edit the article—sometimes, this pattern of checking and editing of one single article happens countless times during a single work shift.

Work Patterns and Orientation Horizons

The above analyses are based on the descriptive data of one action element— the action core itself. However, for detecting work patterns, it is necessary to look at the relations of various elements. Since the action elements have been coded individually, it is possible to visualize their relations using network analysis tools.[11] The network graph (fig. 2) depicts the most important connections between specific actions and resources in the data set.[12]

A very obvious relation shows the basic principle behind this: The action "phoning" (here labeled as Co_phoning_act) and "being phoned" (Co_phoning_pas) are exclusively related to the resource "telephone"; obviously, you cannot do much more with a telephone. The content management system, on the other hand, is the complete opposite: On the given filter level of the network analysis, it is still related to six other nodes. The content management system is not only the most heavily used resource (being part of 14.5% of the activities of an average online journalist), but it is also multifunctional and as such, the center of the journalists' work: It is used for writing and editing various material (both already published web text and original material, which can be written from the ground up—here labeled as: Text_producing—or combined and edited from various sources—here labelled as: Text_change), searching and checking things (Search_own_site) and managing information in

Table 3: *Searching and selecting: share of the activities' cumulated time per person (in percent)*[10]

	NZ A	NZ B	FAZ	SVZ	TS	SPON	Average per journalist
Own website	6.8	9.0	8.7	9.9	7.1	9.4	8.5
WWW. other sites	10.5	9.8	1.5	1.9	2.6	2.0	4.7
News agency client	7.4	6.2	4.0	0.2	9.3	5.0	5.4
Text archive	0.7	-	0.3	0.0	0.7	2.9	0.8
Data archive	2.4	0.2	0.2	0.3	0.7	0.2	0.7
Reading (print)	8.5	12.1	10.3	9.9	13.8	7.6	10.4
Looking through other media	0.1	0.3	3.5	0.2	1.3	2.1	1.3
Press conference	-	2.1	-	-	-	-	0.4
Other	-	-	0.0	0.2	-	0.0	0.0
Sum	36.4	39.7	28.6	22.6	35.5	29,2	32.2

Time basis: NZ A: 36:39:25 h; NZ B: 51:04:00 h; FAZ: 79:57:45 h; SVZ: 95:42:40 h; TS: 77:47:55 h; SPON: 64:44:05 h; Overall: 405:55 h

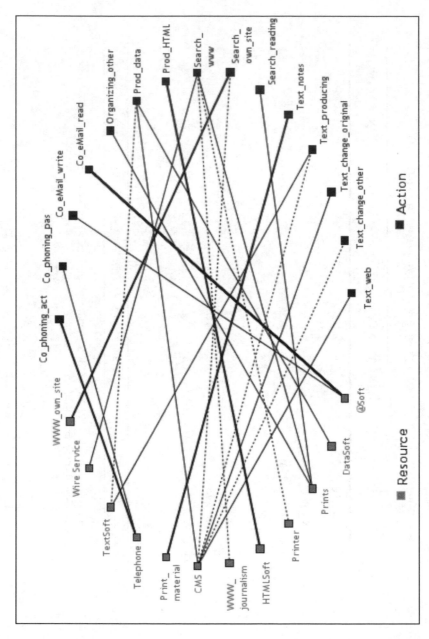

Fig. 2: Network analysis: actions and resources (overall data, relative portion of the relations of the more frequent node; minimum links: 100)

a database (Prod_data), thus being the "Swiss Army knife" of the online journalists. That said, original text production (Text_producing) is done mostly in the word processor (TextSoft), for reasons explained above. Moreover, one set of relations is further relevant to characterize the work of the observed journalists: Internet-based searching (Search_www) is strongly connected to the agency news client (Wire Service) and printouts (Printer). This indicates that the agency system is the primary information source in the observed newsroom (while the link to print outs indicates that news agency material is often printed in parallel to the searching process, so that the journalists have a copy of the most important material for further use).

The network analysis helps to show the relations between two action types. However, with network analysis, even more than two dimensions can be displayed. This leads to complex network charts that can clearly show the structure of the work patterns: In such charts, the nodes are placed closer to each other the stronger the connections are—much like a "gravitational system" (fig. 3). It is interesting to note that this network is purely based on a formal criterion—the association of action elements in one action—but the resulting *action network*, constructed by the everyday work of the journalists, echoes the orientation horizon that the journalists also describe in the interviews.

Central nodes of this network are the topic nodes "national politics" (1) and "international politics" (2), surrounded by text production nodes, searching nodes and data management nodes. This obviously represents the basic journalistic work with the political topics that form the bulk of articles the journalists were writing during that observation. On the left side of the network, however, one finds different nodes: The most significant link is the one between "face to face communication" (3) and "direct colleague" (4). This supports a finding from Altmeppen's observation in radio newsrooms—he calls newsrooms "coordination centres" (Altmeppen, 1999). A lot of the time, the journalists talk to their colleagues to coordinate the production processes, to get additional information on a specific topic or some hint on whom to turn to for information. Other "internal" contacts in the media company are grouped in a circle around the "face to face communication" node. What's completely missing here, are external contacts—because they aren't really playing a role at all. The relation between internal and external communication partners is 8:1, which indicates that online newsrooms are somewhat "closed" social networks. In other words, the journalists in the observed newsrooms are producing news

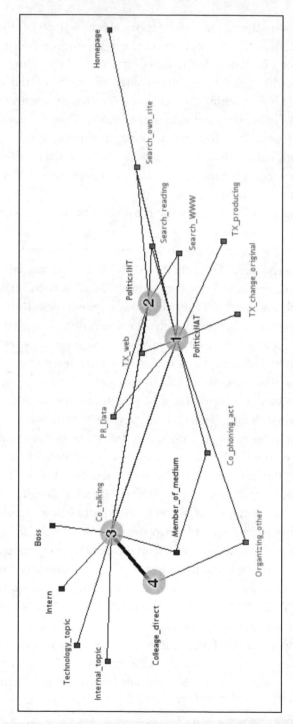

Fig. 3: Network analysis: actions, contact persons and topics (overall data, absolute share, minimum of 500 links)

about "the outside world," but most of the time, they do not make contact with this outside world.

The Outlook: Future Research for Future Journalism(s)

The results of the observation study in five German online newsrooms draws a picture of online journalism that is consistent with other ethnographic studies in this volume while it clearly differs from early visions of a new "multimedia journalism." Certainly, there are some specific "new" qualities of journalistic work in online newsrooms, but overall, the basic orientation pattern is traditional (print/news agency) journalism. What characterizes the actions of the observed journalists is the speed of their work and the dependency on news agency texts.

Technological change still has radically altered the work processes in journalism, though, but probably not in the expected direction. Journalists are not developing into lone publishing stations, running around with headsets, webcams, and phones attached to them—but they use content management systems, agency news clients, word processors, and web browsers much like the elder generation used the phone, the typewriter, and the wire service.

When judging against the early expectations, the observed ultrafast, serial mass production of agency-based news might seem like a disappointment—one might naively ask: Where's the more democratic, independent new way of conveying news in an innovative medium? However, the study clearly shows how necessary it is to observe the social context, instead of being carried away by technological potentials and predicting a media future without doing a "reality check." That does not mean that there are not interesting options for a "different" online journalism; then again, it might not be found in the mainstream media that were observed in this study. It should be noted that this was a deliberate choice: These media are read by hundred of thousands users each day and have a huge impact on the public, so the analysis of their production should be at the core of journalism studies. Nevertheless, independent weblogs, citizen media sites, and podcasts would be interesting choices for future news production studies—they are also sources for (online) journalism these days, they serve as critical observers of journalism, and they inspire journalistic work by doing things differently. However, it might be difficult to conduct "classic" observation studies if there are no persistent physical spaces of news production, i.e., if these media are produced in the spare time of people working together in a virtual network. This is certainly a challenge for future research, together with the constant analysis of new developments in the newsrooms of big companies.[13]

For the German market, the present study is just a first step—hopefully, many others will follow, forming the basis of a more coherent sociology of news work. In times of quickly developing media and changing news market structures, such a critical and independent perspective on journalism is urgently needed.

NOTES

1. In an interview to *Spiegel Online* in 1998, not available online anymore.

2. The theoretical approach cannot be explained in more detail here. See Quandt (2003, 2005, 2007) for a more extensive explanation.

3. Due to work shift restrictions, I had to observe two individuals at the *Netzeitung* in Berlin; a consecutive two-week observation of one individual was not possible there.

4. Using and memorizing codes was also necessary to cope with the work speed of the journalists: A synchronous, full-text description of the journalists' actions would have been impossible because too many things happened in very short time spans. Furthermore, the amount of paper that had to be carried around when following the journalists was minimal, due to the use of a compressed code language and code sheets. By carrying just a pen and a few sheets of paper, it was also possible to walk around without looking strange in the context of a newsroom—basically, all uninformed people (like guests to the newsroom) did not realize that there was an observer present; they perceived the observer as an 'intern'.

5. Excluding about 20 nonjournalistic persons, like webmasters, technical staff, graphic designers, marketing people, etc. In the three previous cases there were also nonjournalistic staff.

6. This is not true for the political department of *Spiegel Online*, where the journalists have very open working times that they organize according to their own responsibility. That said, the main newsroom in Hamburg does have a shift system.

7. This is not always possible for newsrooms that are large and exist for a long time; furthermore, in newsrooms with decades of tradition, the work structures often have been changed to cope with less than ideal configurations of the space, so that after a while, the people do not even realize the problems any more.

8. Central part; the chief editors' room, the office of the editorial assistant and the meeting room are not visible here; they would be found to the left, where the entrance and the restrooms are located. The numbers indicate positions where the observed journalist can be found frequently: (1) observed journalist's work place, (2) newsroom, (B) work place of a

colleague who is a frequent interlocutor, (G) and (D) are work places at the news desk, where the observed journalist also works in late or night shift, (3) is a relatively sound-proof office for telephone interviews.

9. The dependency on news agencies is not only visible by the time used for editing wire texts. If a text is compiled from news agency material, rewritten by a journalist and then changed by her/him later on, this will be counted as changing one's own text (the sources could be still traced back, due to the way how the observation was coded; however, it is impossible to differentiate on such a detailed level in this overview).

10. All types of searching and selection were added to this table, but without indicating special research activities, as research is not an observable category, but the interpretation of an observable activity (like reading something, surfing the Web, or phoning somebody). So you must know the person's reasons to categorize the action as research, and in many cases, the interpretation could be wrong based on imagined reasons rather than facts.

11. Sequence analyses can be done as well, but they are not documented in this publication. See Quandt (2005, 2007) for some examples.

12. The network graph was created using the data mining software SPSS Clementine 7.0. It depicts the elements as nodes, while the links depict how often these elements appear in one action. In the given graph, a minimum number of such links was chosen (100 instances of the respective link) to filter out unimportant relations; furthermore, the strength of the links indicates the relative strength of the relation, i.e., "The links with element B indicate X percent of the overall relations of element A." The higher X is, the stronger the link is displayed in the network graph.

13. There are some signs of a slow change toward more multimedia material and user inclusion (through community-based services and content); however, this evolution of online journalism seems to be a very slow one, especially in Germany (Domingo et al., 2008).

CHAPTER SIX

Maximize the Medium: Assessing Obstacles to Performing Multimedia Journalism in Three U.S. Newsrooms

Jody Brannon

Brannon was privileged to conduct her empirical examination of online news production as an experienced online journalist and survivor of the early days of the online migration of major U.S. media. Her research was oriented toward identifying obstacles toward efficient news production that would maximize the overhyped—but under-realized—potential of online journalism, and she found many. She conveys the pioneering spirit of early online journalism in the most influential U.S. online news organizations, but also the occasionally surprising chaos and poor judgment inherent in their attempts to conquer cyberspace.

EDITORS' NOTE

In digital media, journalists must know when and how to deploy content that takes advantage of the inherent strengths of words, sounds, still images, and moving pictures. Big stories, from terrorist strikes to calamities of nature to elections, for instance, provide the online journalist the chance to showcase multimedia and interactivity, depth and perspective. On the Web, breaking news has become a new kind of service journalism, with digital news teams feeling the responsibility to be reactive like a wire service while pushing to invigorate the most inclusive of media with appropriate methods of engagement and understanding.

In the autumn of 1998, I embarked on field research to assess online news workers' perception of obstacles to presenting online news with a level of sophistication that maximized the medium. I spent a week or more in the online newsrooms of three prominent U.S. news websites representing three traditions of journalism: *USA Today* in Arlington, Virginia, *ABC News* in New York, and *National Public Radio (NPR)* in Washington, D.C.

I brought to my research hopes for blending scholarly rigor with firsthand experience as an online practitioner (I returned to work at washingtonpost.com after completing the research and while writing my dissertation). The field observations were supplemented with 20 in-depth interviews and an anonymous online survey to the journalists in the three online newsrooms (response rate was 24 percent, or 27 responses). Together, these instruments provided a rich understanding of newsroom culture and issues in adoption and execution of news responsibilities from the perspective of front-line workers, senior editors and managers, and executives. In the intervening years of working, teaching and professional interactions, I continue to find validity in my initial scholarly research with conditions in today's online workplace.

My field research, substantiated by subsequent professional observations, informs what I have come to call a theory of "production determinism." Influenced by McLuhan, it accepts the inherent pressures on media output created by technology but incorporates influences of job function and newsroom sociology. For my purposes, I considered many broad theories. The research showed underpinnings of understanding that fused multimedia logic (Altheide & Snow, 1979), technological determinism (Hughes, 1994; McLuhan, 1964; Innis, 1991), and diffusion (Logan, 1995; Eisenhart, 1994; Schreiber, 1977).

A decade ago, online news workers lamented the amount of repurposing their jobs entailed. Many still do, especially those who believe, as I do, that digital journalism can and should be much more than retrofitting the journalism produced for one medium to another. Often, to the disdain of critics and online producers alike, such content online is simply scooped onto the Net, with little regard for context. This is called *shoveling*, an invective that describes people or computers pitching existing content onto the Internet or into carefully crafted presentations suited best for the Web (Martin, 1998; Thalhimer, 1994).

Further, given the nature of the 24/7 medium, the pressures of freshness and updates steal time from creating more interactive content. All too often online producers complain they do little more than repurpose material that a computer has not been nor cannot be programmed to do. This can create a malaise. Recent research has found that online producers often do not add such enhancements (Pedersen, 2006; Chung, 2003; McCombs, 2003; Zingarelli,

2000). The challenge is to put together edited material in a useful, practical, and cost-efficient manner.

This was the challenge that framed my study. Devised to analyze perceptions of technology's impact on online performance, the research identified multiple interdependencies and secondary factors that impede efforts to optimize presentation of content online. Adopting an interpretive approach and comparative case study design, I used a functional analysis framework to assess the extent that technology and other factors influence online news and its practitioners as they applied, adopted or discarded various journalistic techniques on an evolving interactive multiple-medium.

In the course of exploring online media, many news organizations capitalized on their traditional assets. *USA Today* brought to the Internet an experience with and emphasis on words, *NPR* sound, and *ABC* video. Minimally, each continues to bestow a wealth of material on the medium and each attempt to meet the expectations of an audience that gravitates to familiar news organization, given the heritages and styles of news that these media entities represent.

Field Research

The central research question was: What are perceived to be the main obstacles to performing journalism on this medium with greater frequency and consistency in integrating the inherent traits of immediacy, multimedia, and interactivity?

I felt my familiarity with online news operations, combined with my training and belief in qualitative methodology, would answer that question and result in findings that would provide immediate insight and application as well as historical perspective. This approach allowed me to observe the influence of various factors discussed here: technology and production, conceptualizing of news presentations and organizational factors. In evaluating those factors, I then was able to better understand challenges to quality and performing online journalism with greater consistency.

My research was grounded in understanding the journalistic practices and purposes of the three subject organizations. *ABC News* was among the three major network TV news organizations; *USA Today* was the largest-circulating, general-interest newspaper in the United States; and *National Public Radio* was America's broadest radio news organization.

Given that nascent stage of the digital divisions, each was a small version of the larger, established one. Each team seemed shoehorned into available space and staffed more with young outsiders than veterans from the parent organization. *ABC*'s newsroom in New York was largest but seemed the most dense, largely crowded into a partial floor in an older Upper West side building, with broadcast intercoms and editing bays. *USA Today*'s online area had fewer people but the office space seemed more spacious, surrounded by windows on the top floor of the tall building with a tremendous territorial view of Washington. The *NPR* online team was quite small in comparison, with only about a dozen workers, instead of 60 or more, and mostly occupied a few adjacent rooms in *NPR*'s headquarters. Contrary to traditional newsrooms, the online newsrooms were rather quiet. Phones rarely rang and workers tended to stay at their desk and not mill about.

While the average age of workers in the online newsrooms were younger than their traditional counterparts, many of those observed had entered the online arena, leveraging their established journalism careers. Of those working among the three websites, many people hailed from newspapers, the larger industry. A preponderance of employees grounded in words may help explain the reasons so many of them emphasized posting of plain stories and speed ahead of interactivity. They experienced a high frustration level with computers but still exhibited an eagerness to use broader skills than print requires and to incorporate multimedia. In all three newsrooms, workers expressed frustration that they were not staying abreast of techniques and applications that allowed them to be more creative in their work and to strengthen their desire to tailor online content.

Articulation of such frustrations led me to deduce those involved in online news were highly motivated, aggressive, enthusiastic and devoted to doing their jobs well. Frustrations aside, almost all of them were glad their career path had landed them in the online space and they were confident that they were positioned for careers of great promise.

In speaking to the journalists and watching them, it became clear that they felt their success was tightly tied to technology (hardware, software, and production tools) and journalistic skills (packaging and performing). That duality influenced the manner in which they embarked on crafting a multimedia project. However, they also pointed to ancillary factors, including staffing, newsroom management, online style and "best guess" factors that might have been better handled with usability testing or focus groups. As with any journalistic ethnography, workplace issues and newsroom culture encompass the entire study. Organizational structure influenced the way online journalists interacted with technology, production tools, and the product.

Heavy Influence: Technology and Production

One *ABC* producer, night news editor Nancy Trott, discussed in her interview the complicated nature of the many tiers of production (interview, November 30, 1998). A veteran of the Associated Press, Trott admitted that the frequency of production glitches exceeded anything she had experienced at the wire service.

> What I've come to find after getting here is that it's a lot harder to do. There are a lot more technical frustrations to handling the news online than it was with the wire. The focus for the wire was just getting the story written and getting it done. Here it's more getting a story together, into a template, and there's a lot more to it. You can't just say, here's a story, there it goes, it's gone. It's out there, everybody's reading it. Here there have been many, many times when we've had news happening and something back there is not working. And as an editor you feel powerless because you don't know any-thing about what is going on except it's not working. We have this crutch called Big Top, which sometimes works, sometimes doesn't. It's slow. It's difficult to get a wire story up quickly [given] template issues, making sure everything fits, and looks pretty, and that takes time. I think in that way our technology occasionally hampers the way we cover news, or slows us down.

Hardware, essential to fuse Internet technology and technologically con-ceived content, was the third aspect of the question of excellence. As *USA Today* producer David Deal, put it: "A good online service is only as good as its tech department" (interview, November 23, 1998).

The journalists often encountered problems with technology, from their own computers, to the publishing system or applications they used to issues with servers and the Internet. In the course of performing required duties, producers said they used two to eleven or more applications, and the average was six. The learning curve was steep, most people admitted. They acknowl-edged that as time passed, processes improved and frustrations decreased. Journalists' frustration with equipment has been well documented. In general, few among the observed journalists or respondents praised their software-based editing systems. Some acknowledged that tools were growing more stable and sophisticated, at least removing some frustration from the process.

Evolution: Conceptualizing News with New Applications

The perspective of Katherine Dillon, the *ABCNews.com* general manager and vice president, helped to establish the fine line between metamorphosis and sophistication (interview, November 17, 1998). In explaining the evolution of *ABC News*'s kind of news, she discussed the fusion of tools and content. Her comment revealed the overlapping nature of content and technology.

> As our content has evolved, so too has the sophistication of our tools. That boundary is so blurred: where do the tools end, where does editorial end. Everyone here is a journalist, whether they're an engineer or a graphic designer, an audio/video editor or a reporter in the traditional sense. It is quite a good tool—I never really thought I'd hear myself say that. We are fast because of it, but it's a constant process of refining it.

Although I will develop this point of analysis more fully below, it is important to note that Dillon, the online division's top executive, had a vastly different perspective of the tools' abilities than producers who work with them daily. She thought the tool was good and fast; her producers on the front lines of online journalism felt differently. During his interview, developer Alex Storey (interview, October 29, 1998), although clearly frustrated with the production aspects of the job, admitted that *USA Today* was "doing a pretty good job" with available tools and technologies. His words revealed a desire to balance tasks and tools with the responsibilities of online journalism.

> Everybody wants that holy grail, the gizmo that can kick it out before you can even get your finger off the mouse button or the killer app that can allow somebody to customize the site with Java really quickly. People will always be looking for that. In a lot of ways this is just the same old trade with just new toys.

Besides "show, don't tell," most producers, like *ABC*'s Rebecca Leung, said the key to developing online content was to break it down (interview, November 17, 1998). During the conversation, she said,

> You need to start thinking of almost an encyclopedia or magazine, almost a CD [ROM]. You're packaging information. You need to be able to do it so your story stands strong by itself but everything around it adds a different element to it…. You really have to think outside the box.

"Thinking outside the box" was a phrase frequently used in the online industry for getting people to step beyond standard boundaries. Also commonly heard was a reference to people who "get it" or "don't get it," a phrase used to describe those who do or don't maximize the medium and understand the essence of the Web.

One perceived obstacle is time to process news, from recognizing a deserving story, to producing it and posting it online. A common lament was the time-eating nature of production. Said *ABC* producer Terence Nelan (interview, November 13, 1998):

> We're hostage to this production system that requires that we spend much too much time, in my opinion, producing stories, like writing little headlines and clicking to make sure they staged live and everything, stuff like that. The production system is a downside.

Journalists entering or extending the online business at this point may benefit from many lessons learned from other media and experiences, but they also benefit by being able to start with applications that are more robust and easier to use. Not every off-the-shelf product, however, can create an exceptional website, contrary to producers who craved for simple tools to bolster their ability to create content suited for the Internet. *NPR*'s Leon Steinberg said he believed one major reason he had not created better online content was because easy-to-use applications weren't available to help him (interview, October 23, 1998).

> I've been disappointed that I have not been able to do ... this whole interactive, exciting and intellectual kind of presentation. But it has to be simple. And I should think there are software programs that would allow me to put Picture 1 here, Picture 2 there. And I want text on top of it. And I want audio attached to it. I want to be able to establish a discussion with a click, but so far I don't know of anything out there. Maybe it's out there, but I don't know about it.

Frontline workers often are not able to keep abreast of developments in technology in place at other websites, given that each company has different needs, because of existing tool sets. Additionally, given the costs of continuing education or attendance at conferences, online journalists often had little ability to learn about possible applicable solutions to completing their work more easily and with greater confidence.

ABC homepage editor Adam Glenn, said such projects conveyed to him the necessity of presenting information in new ways (interview, November 12, 1998).

> I think if you bring a mind-set, you're already in a hole, other than in the most obvious journalistic creed of making sure that what you say or convey is truthful and fair. But anything other than that I think is a setback. Most of the stuff I tried to force myself to do successfully—and unsuccessfully—as a producer was to stop thinking the way I did think about presenting information.

Like others interviewed, Glenn's statement captured many of the factors perceived as the main issues in the conceptualization process of online content. Staff members were adjusting to a new medium. They were from various media backgrounds. Many were inexperienced at any medium, much less this fledgling one. They used different tools, requiring a broader way of thinking. They felt the pressure of immediacy and often failed to take further steps to develop content suitable for the medium. They appeared to be mired in a cut-and-paste rush where volume was easier to produce than quality work.

Relationships: Organizational Impact on Content Development

Problems with staff and resources were frequently cited as impacting the way news was handled. Managers at all three companies mentioned the difficulties in exerting discipline on staff, in trying to focus updating and development efforts that best benefit users and the website. While they looked to larger teams to build deeper, fresher content, they said staff alone would not resolve issues of content presentation. *USA Today* publisher Lorraine Cichowski said challenges of management, training, and growth created "people issues [that] never go away" (interview, October 30, 1998).

> We have really three levels of staff. We've got the senior people, the grizzled types who know management and know journalism. We have a lot of entry-level people who are just brand-new reporters, baby reporters. And so it's just…teaching them what the standards are, and it's the same challenge you see in any local newsrooms, not really at *USA Today* but elsewhere. And then the middle level is your line managers, and almost [all] of them are in their first management job. So you were teaching them not only the journalism skills you wanted but you're teaching them how to teach and grow and manage. And in the meantime, shuttle launches are going up and scandals are happening.

Given the need to continually exert news judgment, most people rated that skill as the most important, even ahead of writing skills.

"Writing for the eye definitely is different than for the ear," *NPR* writer Cara Fogerty, said in an aside while I observed. At *USA Today*, the differences between people hired online versus in the main newspaper operation were stark. Explained Harvey Laney, *USA Today*'s news director (interview, October 30, 1998), who agreed with colleagues that online editorial specialists needed to know more technology and production than their print counterparts:

> I like everyone to be utility infielders, so that no matter who is here or who's working, the job is going to get done. Everybody knows how to do everything. Everybody is

capable of doing everything. Some are better than others. Some may be better writers. Some may be better stronger technically, with tables or whatever, but everybody has the ability to do everything. There is no specialization, whereas downstairs you've got a guy who just writes headlines all day long, so he spends his whole time trying to write a decent, award-winning headline or a guy downstairs who just writes about the Supreme Court, so he's very well versed at that. We don't do that. We expect people to have a broad overall knowledge base. So next year, there is an era of generalization coming in. I think it's good in terms of that it makes people interchangeable when you're looking at scheduling or you're looking at what needs to be covered when by breaking news. I think it's bad because you don't have the expertise that you'd like to have sometimes.

Challenges to Quality

In analyzing responses, I clearly discerned that both managers and producers perceived technology to be a culprit in their respective quests for quality and satisfaction. My interpretation of responses showed that managers lamented the content for not being more medium-appropriate because news workers got too bogged down in the production process to think broadly or creatively; producers, meanwhile, felt their tools were so inadequate and time consuming that the workload precluded them from developing content better suited for the Internet.

The range of answers to the issue of quality was wide. In the field interviews, subjects often pointed to similar issues that acted as impediments to quality, specifically budget restrictions, staffing shortfalls and turnover, production obstacles and experience from various media. Listening to their perspectives, I believe it was clear that most people saw technology or technologically dependent production tools as being the biggest obstacle to consistency. Too often, the process impeded efforts to excel. Perhaps with a nod to an *NPR* editor who admitted going "numb" updating, too weary to then create an interactive content element, *ABC*'s executive producer Mary Bruno said, "It's hard to be a careful editor when you're shoveling coal all day" (interview, November 17, 1998). Web content may not necessarily be evergreen, but much of it was always ready to be accessed, and most online reporters disliked their updating duties.

Obstacles to Consistent Online Performance

In arriving at this point, we have seen, in the words of those people on the front lines of inventive journalism, evidence of occasional chaotic conditions and uncertainties. Rapid change and rampant desire to more fully embrace the Internet were complicated by uncertainty and inexperience. Many people said the online industry has progressed quickly.

In the weeks I spent observing the front lines of pioneering digital journalism, it was apparent that the days were exciting yet arduous. Very little was routine. Five of the 20 people I formally interviewed at length specifically stated that there were no rules in online journalism, a position supported by the recurrent perception that dexterity was the most important trait a cyberspace worker should possess. More than half of the interview respondents directly mentioned versatility as a quality important to online journalists.

Young managers and morphing methods of performance created anxiety for producers; managers and executives, meanwhile, experienced nearly 100 percent turnover in certain departments and acknowledged the obvious: "The quality of your employees will determine to a degree your success," USAToday.com's editor Jim Schulte said (interview, October 30, 1998).

Based on themes observed in the online newsrooms and comments from interview subjects and survey respondents, I concluded that successful online journalists have embraced the basic tenets observed by their traditional-media brothers and sisters: news judgment, accuracy, fairness, and basic communication competency (i.e., good writing, clear articulation, solid reporting). Many of the characteristics perceived as important to an online journalist were not unique to new media but perhaps were necessary to a deeper extent: the ability to multitask, comfort with technology, tolerance of uncertainty and ambiguity, and, ultimately, as one survey respondent said, possession or development of "the ability to tell stories and make them come alive in ways you can't in other [media]."

Recall the perspective shared by *ABC*'s Glenn, who said he continued to meet greater success when he managed to ignore, to a certain extent, the media mind-set he had developed over years in the news magazine business.

> I'd say the thing I've really had to shed here is a feeling that the word is primary. I don't think it is. I'd probably have to say, to be honest, many of the folks in the newsroom, particularly the producers, are still cleaved to the idea that the word is not only the most important thing but in many cases the only thing. That is very one-dimensional in a kind of three-dimensional realm. If you focus just on what you write and don't think about the various ways in which words can be presented, much let alone how to step aside from words and use images, animation, interactivity, to make a story come alive, then producers are not really taking advantage of the medium. And frankly we do it successfully many times, but many times we don't.

At all three sites, news workers complained about the rote duties and time expenditure of processing news. *USA Today* producer Alex Storey made the process of creating online polls sound like easy drudgery:

> This is kind of plug and chug. It's like we're generators. Type in your text, pick your colors, send it out, and you've got something added to the site, people will interact with it. And it's not too much trouble.

At *USA Today*, Laney's admonition to "stop churning out briefs, start churning out packages," the news workers clearly recognized the tone set by management. The dichotomy of what people were told versus what they did was striking. Producer David Deal said,

> I can make a beautiful package if I have an hour to do it [but] I have 10 minutes, and what I can do is run a brief on it, look for another website, add a picture to it.

This package was the minimum: Laney would say to Deal that a package could be simply embedding a photo. In fact, I detected a feeling at *USA Today* and elsewhere that people either didn't have the time or didn't take the time to think of ways to present content more suitably for the Web.

Based on my findings in this comparative case study, I define three keys to creating online journalism with greater consistency and more interactivity. One is training, or retraining, as *ABC News* executive producer Mary Bruno said, which seemed necessary in a medium with evolving tools, techniques, and strategies. Another is mind-set, or developing and nurturing the characteristics of the technologic journalist. The third is recognition of the trade-offs in quality and job satisfaction that accompany an emphasis on the Internet's easiest and most visible inherent trait: immediacy. As one *ABC* respondent wrote, "Speed comes at a price and quality; clarity and creativity are oftentimes the currency."

Conclusion and Implications

The research assessed obstacles to routinization and more sophisticated presentation of content. My analysis showed that online news teams had difficulty in adjusting to technologies and demands created by the need for multimedia thinking. Reasons ranged from too few staff to workers ill-equipped (by training/experience or by hardware/software) to market and organizational challenges.

The cumulative data from, field observations, and interviews, complemented with the survey, showed various impediments, varying in import and intensity, to performing multimedia journalism with consistency. In assessing chaotic conditions faced by news workers on an evolving medium, the comparative case study approach helped to distill the data to elements essential to maximization of the potential of online journalism.

This qualitative analysis showed what online journalists faced, as the 21st century loomed. It showed that, regardless of tradition of service, online news teams from newspaper, radio, and television backgrounds shared similar challenges: underdeveloped technology, inefficient production tools, limited and often young staffs, bureaucratic shifts, young managers, and large ambition.

In allowing early impressions to be recorded, the perspectives of online workers showed how the early online workers tried to make sense of virtual journalism. Subjects shared their own opinions in interviews or in response to the survey question, "Working in cyberspace is…"

> "More stressful, but (occasionally) more rewarding. Very frustrating, but also very satisfying." (*ABC* respondent)
>
> "Challenging, tiring, and often satisfying." (*USA Today* worker)
>
> "Difficult, cool, and rewarding." (*NPR* staffer)

A new kind of journalism cannot be created unless the tools are in place, but also imperative is a richer understanding and development of people who operate those tools and conceptualize online news. As *ABC*'s general manager Kathleen Dillon said:

> The technology is fairly crude. The software and systems are user unfriendly and prone to break. We have two teams of people who do nothing but deal with technical problems. That's an unfortunate waste of resources but it speaks to the cutting-edge nature of the medium. This is an industry in its infancy, and our tools reflect that.

Key to my argument is the degree with which someone's work reflects a digitally sharp mind-set, able to adapt to using tools still considered crude by most. Only people who regularly exhibit it tend to maximize the medium. This requires both discipline and instinct. It is the skill of "originality" that *ABC*'s Mary Bruno said set online editors apart, the "initial burst" that *USAToday.com* deputy editor Chris Fruitrich said is needed to follow the first push at a breaking story and the avoidance of the single dimensionality that *ABC*'s Adam Glenn described.

An online journalist has an arsenal of ways to convey a news story, given that the Internet is innately multimedia. Basically, I extrapolated David Altheide and Robert Snow's theory of media logic (1979) as multimedia logic—a mode of interactive thinking that online journalists are learning to apply when packaging news on a maturing multimedia environment. The theories of determinism, diffusion and logic stimulated further thinking about their nexus.

The truly "technologic journalist," I suggest, possesses this digital mind-set, unobstructed by unreliable or clumsy tools of production. I believe my findings show likelihood that online journalists who are adaptable, versatile, technologically adept, and patient with the corporate, market, and Internet maturation processes are most likely to be satisfied in their job. In turn, they are reflective of a new breed of professional journalist, one that is more likely to create content that maximizes the medium and helps to usher an industry toward convergence—and one that embraces and celebrates technological possibility and editorial excellence.

CHAPTER SEVEN

When Immediacy Rules: Online Journalism Models in Four Catalan Online Newsrooms

David Domingo

Different media contexts can foster divergent developments and definitions of online journalism. The working routines and values in the online newsrooms of two newspapers, a broadcaster, and a stand-alone news website in Catalonia (Spain) were examined to identify similarities and differences. Immediacy was the main motivation for those online newsrooms embedded in traditional media companies. Interactive features were viewed as a problem rather than an opportunity in those newsrooms, while the online-only venture was more eager to explore audience participation. However, in all the cases, the professional culture of traditional journalism greatly shaped online working routines.

EDITORS' NOTE

Retracing the factors that played a role in the development of online journalism is an exciting challenge. The narrations of those involved in the early stages of the projects and the unsolved conflicts surfacing in the daily routines give worthy hints that an ethnographer can follow to rebuild the complex history of strategies, constraints, and values that shaped a news website to be what it is. This chapter[1] discusses the implications that the social (professional culture, newsroom organization) and material (technical resources and skills) contexts of four different online journalism projects in Catalonia (northeast of Spain) had to their development. The analysis takes a comparative perspective, based in case studies of online newsrooms in different media environments: broadcasters, newspapers, and purely digital outlets. The main hypothesis underlying this approach is that media traditions may be one of the crucial factors that shape Internet adoption and use in news media companies. By media traditions I mean journalistic values, routines and product formats (Altheide & Snow, 1991), the ways in which a news product is produced, which vary from company to company in detail, but are clearly standardized at large in the different traditional media (TV, radio, newspapers) as the sociology of newsmaking has shown (Manning, 2001; Schudson, 2003).

The aim of this approach is to build a comprehensive and nondeterministic description of the directions which online journalism is going.[2] The ideal models proposed for online journalism in the 1990s tended to forget that any technological innovation is adopted locally and actors make decisions based on general trends but also local constraints and assumptions (Boczkowski, 2004b). There can be more than one model of online journalism and empirical research suggests that the ideals of the 1990s, despite being a crucial referent for Internet journalists, cannot predict the path of online news evolution.

Case Studies and Methodology

Catalonia has a semiautonomous media market within Spain. Catalan-language media on the Internet form a cultural communication space (Moragas, Domingo, & López, 2002) that became explicitly visible when the ICANN approved the .cat web domain for websites in Catalan or about Catalonia. Among general interest online news sites in Catalonia, four cases were chosen for the research, with the criteria of selecting different media traditions (print, broadcast, and online-only journalism) in order to compare their similarities and differences to test the diversity principle of the theoretical framework:

- A purely online project: *laMalla.net* (www.lamalla.net)
- A newspaper online venture: *El Periódico de Catalunya* (www.elperiodico.cat)
- A broadcaster's online portal: *3cat24* (www.3cat24.cat)
- A local online newspaper: *Diari de Tarragona* (www.diaridetarragona.com)

Work routines, professional values, and innovation strategies were the main aspects analyzed in each case. Observation of the online journalists at work was conducted in 2003, consisting in five stages over the course of three days in each of the four newsrooms. A weekly rotation allowed the researcher to visit every media company from month to month, and this helped to detect more easily if the product or the routines were evolving. Informal conversations were undertaken in the newsroom to make explicit journalist's definitions of their work and the technologies they used. During this phase, documents defining the news websites and routines were retrieved in the field, especially those emanating from the first stages of the projects. In a second phase which occurred in 2004, 21 in-depth interviews where conducted with people associated with the production of the news websites in the past and present: editors, reporters, technical managers, and traditional media managers. They were asked to reconstruct the evolution of the project from their point of view, their definitions of online journalism in general and their particular online venture. Follow-up visits were paid to each online newsroom in 2006, and new informal interviews were conducted.

Immediacy and Other Online Journalism Utopias in the Work Routines of the Newsrooms

The online journalism *utopias* developed in the 1990s (from interactivity to multimedia, from immediacy to the bottomless newshole) were shared by professionals in the analyzed newsrooms as assumed ideals that came up in conversations. They felt the need to justify the technological choices that did not fit the ideal model by blaming the limitations imposed by the social, economic and technical context of their media company. Many times, their vision of the future evolution of their website was close to the ideal model, while the present was defined as a limited version of what they would like to develop.

But not all the utopias had the same importance for these online journalists. Immediacy was clearly their main choice, as found by other authors in this book. The ability to permanently update the website was seen as the defining feature of online news when compared to traditional media, even if 24-hour television news channels have already been around for years (Kansas & Gitlin, 1999). The primacy of immediacy was overwhelming in online newsrooms dependent of traditional media. At *laMalla.net*, the online-only project, reporters had assumed they could not compete with the bigger media outlets and only applied the immediacy rule in the biggest of breaking news. They tried to develop an alternative news agenda to attract their own public and produced those stories without the pressure of time. While in the former cases the main aim was disseminating information quickly, as reported in German and Flemish online journalists' surveys (Quandt et al., 2003; Paulussen, 2004), at *laMalla.net* it was interpreting current events and offering useful services to the audience, as suggested in the Dutch online journalists profile reported by Deuze and Dimoudi (2002).

Most of the work routines in traditional media online newsrooms were shaped around the choice of immediacy. Journalists tried to publish a story as soon as possible and news agencies wire services were the perfect source for that purpose. Depending on the size of the newsroom (some had just one reporter per shift, others up to three or four), wire texts were edited or not before publication. *3cat24* had the most complex production organization, as reporters had access to the broadcasting newsroom texts as another news material input besides the wire stories and a linguist[3] and an online editor oversaw the stories created by the reporters before publication. At the newspapers, online reporters (no more than two per shift) self-published the stories at the homepage and decided the hierarchy on their own. Print editors advised them in case of doubt, but no systematic supervision was done on website news.

LaMalla.net reporters also had a self-publishing strategy, but the reason behind this parallel routine to online newspapers had a different rationale: every reporter had specialized in two or three topics and managed the sections of the site autonomously, putting their stories onto the homepage. The online director oversaw the process and set common criteria. Specialization let reporters build a stable group of online sources that made them less dependent on news agency material. They called firsthand sources on the phone and went out to the street to cover a story, not as often as print or broadcast reporters, but much more than traditional media online journalists, for whom almost the only source besides wire stories were their offline newsroom colleagues. In fact, this would only happen in very rare occasions, when online journalists doubted seriously about a wire story and decided to ask an offline counterpart about the story. Many online reporters felt like "second-class citizens," as Singer, Tharp and Haruta (1999) already detected in U.S. online newspapers and García notes about Argentinean journalists in chapter 4.

The killing of a Spanish journalist by a U.S. tank shot aimed at the Palestine Hotel in Baghdad in the early stages of the 2003 invasion of Iraq was the most dramatic moment during my observation stage at the *3cat24* online newsroom. It is an example of standard work routines being disrupted by an event that challenges the usual news production process in the online newsroom. The Palestine Hotel was well known during the Iraq war, because it was the place were international journalists stayed and the place where live stand-ups were recorded when the U.S. soldiers started laying siege to Baghdad. That morning, April 8, 2003, around 11 a.m., amidst the uninterrupted wave of news agency material about the situation in Iraq, Reuters was the first to inform the world that the Palestine Hotel had been attacked. The first wire story was just a headline. After some minutes, EFE, the leading Spanish news agency, sent a paragraph about the tragedy: two reporters were wounded after a tank shell hit the hotel. *3cat24*'s online subeditor (the person in charge of editorial decisions on routine days) considered that it was enough information to produce an initial piece of news. When she was editing the lead, before publishing it, a new wire story came in announcing that one of the journalists was Spanish. She changed the headline to reflect this information and added it to the lead. Once the story was ready to be corrected by the linguist, the subeditor printed the new wire stories (there was new data coming in almost every minute) and prepared a longer text. A reporter shouted that *El Mundo*'s website (the leading online newspaper in Spain) was reporting five journalists had been wounded. The subeditor felt she'd better talk to the editor, usually not involved in breaking news decisions, to know his opinion regarding where to publish the story on the homepage. He suggested substituting the story at the second position (*Al-Jazeera* journalist killed some hours before) and include that old story in the text of the new piece. The subeditor worked on merging the pieces and updated the number of wounded journalists. The reporter found a new wire story stating that the Spanish journalist had lost a leg. Meanwhile, the first news agency photos were getting in from the EFE online service the newsroom had recently subscribed to. They decided to reposition the piece in the right column of the homepage, the one reserved for stories with good photos, and posted a picture of the wounded man. After reorganizing the homepage, the subeditor took a wire text from Europa Press (the second largest news agency in Spain) written in Catalan (the language of their news portal) to add details to the story quicker.

After a while, they realized that radio texts (of sister station *Catalunya Informació*, which they could access anytime) reported that one Reuters journalist had died. They looked for wire stories saying so, but they did not find any. They decided to wait before editing the piece, arguing that they knew the radio station did not have any correspondent in Baghdad. After a while *El Mundo* stated the same, quoting *Al-Jazeera*, and EFE confirmed it as well. The editor argued they should wait for Reuters to confirm the fact, as the journalist worked for that news agency. The subeditor decided to walk downstairs to the television newsroom and ask them. They told her that their correspondents in Baghdad had seen the dead journalist with their own eyes and when she went back to the online newsroom she quoted them as the source in her piece.

Everyone was so focused on that story that they forgot that they wanted to update the first one (on the Baghdad siege) and also to edit a story on the use of chemical weapons by the Iraqi army that had later been denied. The subeditor was already back in the newsroom when a wire story suggested that a U.S. tank fired the shell into the hotel, but it took an hour before she realized that that data was missing from her story. She decided to start anew, and took the piece of *El Mundo* and the last radio text as the base to rewrite her story. Soon after she finished, the reporter highlighted a wire text stating that the Spanish journalist was going to survive the wounds. One minute later, she saw a new wire story reporting he was dead. The online team was in shock after this contradictory information. They kept on asking one another, "Does any other wire story say something else?" The subeditor edited the news story again with this latest data, but she decided not to publish it until another source confirmed it. The editor phoned the TV newsroom and they said they could not confirm the death of the Spanish journalist yet. Thus, the editor suggested watching *Telecinco*, the Spanish channel for which the journalist worked. "If they do not say it, we can't," he argued. There were five minutes of tense waiting. Then, they received a call from the TV newsroom confirming the death; *Telecinco* soon interrupted their programming to state the fatal ending. Before publishing the last version of the story, the subeditor edited the first paragraph four times, not satisfied with the wording. Everyone in the newsroom was very touched, sad, silent.

This was not a common story. It was the story of the murder of a journalist, and the reporters at the online newsroom witnessed it live through the wire service. They saw him go into the hospital, they heard the doctors fighting for his life, they felt the hope, and they wrote about his death. This was one of the few occasions when the news in the wire service had to wait before being published; who had been killed was one of them, a journalist. There was no rush, time had stopped and online reporters could not believe the wire stories coming in. An infinite sadness broke the routine.

Nonetheless, the rest of the story describes perfectly the newsgathering work in online newsrooms of traditional media, resonating with Quandt's findings in chapter 5: constant incoming wires, rewriting work with the news stories, competitors as an essential reference, traditional newsroom as an exceptional source. The daily routines of online journalists are more concerned with filtering and selecting a constant flux of information rather than an active search for stories.

Striving for immediacy also had important consequences on the develop-ment of the other utopias. Among other factors, immediacy displaced the ideal of in-depth reporting from the daily routines, and made it very difficult for traditional online media to deal with interactivity. While all the online websites had user forums, journalists did not use them as sources for story ideas or to better understand the needs their audience. Reading the forum posts was a routinized activity done by one person per shift in order to delete inappropriate or offensive content. Journalists argued that they did not have time to pay attention to users' feedback. Instead, at *laMalla.net* they had been more perme-able to the utopia of interactivity; beneath each news story there was a space for user comments where reporters also wrote to greet users' suggestions and opinions. A "representation of users" (Boczkowski, 2004b, p. 199) based on the ideal of an active Internet user instead of the mass model of traditional media inspired the attitude toward interactivity of the online-only news site.

Multimedia development was limited in all the cases. *3cat24* and *laMalla.net* published daily audiovisual material repurposed from traditional media. The newspapers did not have any in-house sources for such materials, even though their web servers were technically ready for webcasting. Both newspapers used to have infographics, but production priorities forced discontinuation of their publication. Online editors argued that multimedia content was still not in demand by their audience. Broadband connections in Catalonia grew fast in the early 2000s, but still half of Internet users have slow modem connections that make streaming or downloading a video a painful experience. This was one argument to justify the scarce use of multimedia content, but editors also believed that simply repurposing television content might not be attractive for users.

Special coverage of events was the "innovation edge" of online news projects, the occasion for pushing the envelope and exploring the edges of online journalism that daily routines did not allow for consideration. Being outside of the daily rhythm and mostly devoted to planned events, special coverage let online journalists think ahead in features and concepts. The other facilitator of these innovation edges was the fact that they were necessarily short-term projects with a limited scope. While general website development was an open, never-ending process that could be delayed if necessary, specials had to be ready when the event was scheduled. They were the space for utopian experimentation: participatory publishing where users could become content producers, multimedia-rich reports, and complex hypertext structures with in-depth background on an issue. In a way, specials were the institutionalization of utopias: what daily routines could not handle, specials offered a routinized way to develop it.

Traditional Journalism as the Social Context of Online News Production

Online news production routines in the analyzed newsrooms reproduced the gatekeeper model of traditional journalism, without challenging any of the existing journalistic culture and even lacking some usual tenets such as fact-checking. Thus, in these cases, online journalism has not revolutionized the profession, but developed as a new specific form of journalism. The reason for this continuity of online journalism has to be found in the mind-set of the members of online newsrooms rather and the lack of support from the management of their media companies for a more exploratory attitude toward the Internet. Even *laMalla.net* mostly reproduced traditional journalistic roles to gain a status among the profession. There was neither direct involvement of traditional newsrooms in online projects' concrete development, nor in daily production. Even though online newsrooms' relative position inside media companies in terms of physical location and organizational links to offline newsrooms varied among projects and over time, this was not a determining factor in shaping online news production routines, contrasting with Boczkowski's (2004a) findings when comparing three very different projects of U.S. online newspapers.

Online and traditional journalists tended to mutually ignore each other in most of their routines. Online newsrooms had the perception that the Internet project was not important for the traditional media staff. They argued that there was a general lack of knowledge about the Internet as a news medium in traditional newsrooms, but they mainly blamed the lack of economic autonomy of online projects (as they were not profitable businesses) and that the websites were seen as competition for the traditional outlet, especially in newspapers. No original reporting from the newspapers could be scooped online. Collaboration between both newsrooms was seen as positive, but it was difficult to incorporate it into the daily practices. Their own news production processes absorbed the energy of each newsroom and few practical collaboration initiatives were explored and maintained afterwards.

Even though online journalists felt that they had won credit among their traditional counterparts after big news events such as the Iraq war or the March 2004 bombing in Madrid, these signs had not changed the daily routines of each newsroom and both products lived parallel lives, without much overall coordination. There were some consultations, traditional journalists advising online journalists, but these limited contacts could be regarded as exceptions, as the online production rhythm did not allow for consistent interaction. Online journalists were the ones who had to take the initiative for this to happen. In content coordination this was also the case: it was online journalists who tried to make sure that their product was coherent with the editorial line of the traditional medium. Traditional journalists seldom visited the websites.

With such poor daily relationships between newsrooms and given the relatively high autonomy of the online projects in traditional media, it does not seem plausible that offline newsrooms played an important role in directly shaping online journalism routines. However, we may argue that the very attitude of the traditional newsroom management neglect of the online project was actually a crucial shaping force. Online directors and reporters, in these cases, often referred to the lack of understanding and support from the company managers, and more concretely to the impossibility of asking them for more human or material resources, as the main constraint in the development of their online news outlet. Online newsrooms were conscious that they were the last priority of the company and could not grow beyond the existing limits. Therefore, traditional journalistic routines and values were not an imposition of the offline newsrooms, but something deeply rooted in the mindset of online journalists that was put into dialogue with online journalism utopias to shape the online production cultures. The choice of immediacy as the main value, following a general trend in journalism as a whole, and the limited resources delineated the actual development of the rest of the utopias.

Technology as the Material Context of Online Journalism

Technical artifacts can be regarded as an active actor in the shaping of a technological system (Callon, 1987). They may be conceptualized as the material element in the system, the one that can be reduced to hardware and software tools that online journalists use in their daily routines. Obviously, the features of these technical tools have been socially defined. Some of them were devised outside the newsroom, by companies specializing in them, but there was a choice of a concrete solution by newsroom managers, based on three aspects: their expectations about what that software and hardware should be used for, the current technological options available and the budget they had allocated to each tool depending on the priority it had in the production process. Besides this, web content management systems (CMS) were designed in-house in most of the cases, and therefore we may infer that they were devised to perfectly fit the needs of the online newsrooms, their definitions of the product and their routines, although this was not entirely the case.

Technology was not an external input to online newsrooms. The material context of the journalistic work was the result of many technical design decisions that clearly affected the performance of reporters. The technical department in the media companies and its relationship with the online newsroom played an important role in these decisions. Thus, the material context was socially shaped, but once defined, its materiality actually became a crucial factor in shaping work routines. Reporters usually did not have the chance to participate in technological decisions and one of the strongest internal social conflicts in the newsrooms arose because of the frustration with the technical features of the tools they used, as the work of Brannon (in this book), McCombs (2003), and Boczkowski (2004a) have also highlighted. CMS design did not always fit the needs of journalists, and discouraged them from routines that would have sufficed in other material conditions. In other cases, they complained that technical routines were too cumbersome and time consuming, working against their wish for immediacy. This led to a relationship of distrust between the journalists and the CMS staff.

The four newsrooms had similar CMS, but they were used differently, having varying routines for news writing, link management, and homepage updating. Even within the same newsroom, there was not a unique way of performing CMS routines: every journalist had his/her own solution to deal with technical tools. I may argue, then, that technological tools did not predefine work routines. Rather, the CMS materialized the main production routines and workflows in specific tasks on a web-based interface. Variations in performing technical tasks are a symptom that there was no systematic training of reporters in the newsrooms, but also that technical tools do not predefine a set of uses and routines.

Journalists in the different online newsrooms had varying technical skills. Some had advanced HTML knowledge, but this was not relevant to the quality of the online product. At *3cat24* technical staff tried to develop automation technologies to lower the technical skill required of journalists to perform any routine. In other newsrooms, part or all of the reporters performed a great deal of technical work (encoding video, editing an HTML layout, editing a picture for online publication) with diverse results in product complexity.

None of the analyzed online newsrooms deemed their relationship with technical staff as ideal. The general feeling of journalists was that technical staff was far away, slow or even deaf in regard to solving their problems or responding to their needs. The technical staffs were too small to deal with long-term development and, short-term problem solving and coding specific solutions for special feature articles with the timing that journalists would like. Technically skilled journalists in the team (at the side of the online newsroom or of the technical staff) were crucial in all the cases, to communicate journalistic needs to the programmers. However, usually these technically skilled journalists were not involved in daily production and the needs of breaking news production were not fully taken into account.

This is why technical tools evolved quite slowly in the eyes of the journalists. CMS always had noncritical bugs that took months to be solved, in the queue for the next version. They made journalists life less easy, but they developed routines to deal with the bugs. Audiovisual production tools evolved faster as they were innovations for the whole of the media companies, but there were always drawbacks to fit the needs of the online journalists. Even though online reporters did not usually participate in future project designs, they indirectly participated in innovation processes by accepting, rejecting, reinventing or adapting new technical tools or work routines. They applied the logic of existing routines and online news values to try and make sense of new developments. If new developments were coherent with those existing rules they were assumed without trouble, but if not, they might be rejected. As I have shown, sometimes rejection was as easy as journalists ignoring the new feature.

The Dynamics of Project Development

Projects were constantly and slowly evolving, but the fact is that the core philosophy about the product and the production culture changed very little in all the projects during 2000–2006, regardless of the changes in location, design, CMS and technical tools. The four models analyzed seemed to have some sort of stability. This might not be considered as a proof of maturation, but rather as the effect of inertia, the difficulty of reshaping a process when a product has to be constantly updated. Online directors and journalists had very clear ideas of the shortcomings of the project and the ideal routines and product concepts they wanted to work toward. There was a will for change and a perception of being in a learning process that was ongoing.

Online newsrooms borrowed the software "version" metaphor to deal with constant evolution in an intensive daily work oriented project. They accumulated possible new features and accelerated structural development for some time, fighting against short-term needs. Usually technical and design changes were the easier ones to make and to see, but even these could not be fully developed following an initial plan. Organizational changes were much more difficult to put into practice. This is the paradox of materiality: it takes lots of hours and resources to develop a new CMS feature, but it is easier to make it concrete and to plan it. Organizational changes, especially if they pretend to involve the traditional newsroom, were much more difficult to manage. Social attitudes were harder to shape than software to develop.

New versions of a project, usually focused on technical web development, better enabled social change if organizational innovations had matured for some time. But even in this context, change was not easily undertaken. While online journalists tended to see technology as a brake for development of their ideals, many times organizational and technical skills provoked dysfunctions in the innovation processes. From the perspective of online journalists, project evolution was mainly perceived as technology driven. The journalistic or strategic decisions underneath these changes were transparent to them, and most reporters felt they were mere spectators of innovation processes in the project. In this context of slow permanent change, where a new version was foreseen for months or even years, online newsrooms had a sense of being in a constant temporary situation. This helped journalists in adapting to innovation, but also promoted a passive attitude, where they would rather wait for a better future than actively reflect on the present limitations to overcome them.

Conclusions

The four case studies of this research clearly suggest that there are multiple factors that shape the use of the Internet as a news publication channel. They also highlight that there are common values and routines in online newsrooms and at the same time important differences. Each context (each company, in this case) develops concrete strategies, definitions, tools, routines, and roles that can only be explained by a deep analysis of the actors and material conditions in each environment.

The law of radical potential suppression (Winston, 1998) can interpret the process of the adoption of the Internet as a news medium in the four online newsrooms. Social groups react to innovation by adapting new technologies to the established rules in order to prevent radical change, which could be dangerous to the structure of the group itself. While utopias and competition acted as the main accelerators of the phenomenon of online journalism, traditional journalistic culture and the scarce human and technical resources (due to the lack of trust of the new medium by the companies' management) acted as brakes of innovation. At *laMalla.net*, where the traditional journalistic culture was not as institutionalized as in the other online newsrooms (which were semi-autonomous annexes to traditional newsrooms), there was evidence of a more original model of online journalism: utopias were more clearly developed and their online news product was shaped with a sharper contrast to traditional journalism. However, the striving for recognition among professionals also made them reproduce most of the traditional routines, and the shortage of resources left them far from some of their goals.

Further research (and this book is a first opportunity) should compare more case studies across different countries and industries in order to find general trends and variations. This would allow online journalists to understand the role of utopias in their daily routines and the material and social constraints that shape their definitions and strategies of online news production. This way, they would be empowered to critically reflect on their work and take informed decisions when exploring the opportunities of the Internet as a news medium. There might be many ways of doing online journalism and the constraints of any given media company could be reinterpreted as a framework within which there could be different options to improve the quality of the online news product. Researchers may contribute useful advice in this process of redefinition of online projects. Active involvement in media design and development would be beneficial for both sides (Castelló & Domingo, 2004). Internet users' habits as consumers and producers of online information are a mostly unexplored research area that can also offer fruitful complementary data for researchers and for media professionals.

NOTES

1. A first version of this chapter was presented as a paper at the COST A20 Conference *The Impact of the Internet in European Mass Media* (Delphi, Greece, April 2006).

2. See chapter 1 for the theoretical framework of this approach.

3. CCMA, the public broadcasting company managing 3cat24 is owned by the Catalan government and has the duty to preserve and promote the quality of Catalan language. As the role of linguists was already part of the radio and television newsrooms, it was natural for them to add it also to the online newsroom.

CHAPTER EIGHT

Online Journalism in China: Constrained by Politics, Spirited by Public Nationalism

Johan Lagerkvist

News production in China receives a great deal of interest from journalism scholars and political commentators alike, but to date there is little evidence of comprehensive ethnographic research into any form of Chinese journalism. The closed nature of Chinese society and traditional subservience of journalists to the Communist Party may account for this, but so too does the only recent acceptance in China of Western techniques of media research accompanied by an increasing openness to Western researchers. Through his interview-based research, Lagerkvist reveals much about the changing nature of online journalism within the contradictions of China's recent history.

EDITORS' NOTE

With the countdown to the 29th Olympic Games in Beijing in 2008, the condemnations from human rights organizations of the Chinese regime's handling of dissidents and free speech increased dramatically. This was despite the fact that the central government in Beijing had softened its dictatorial attitude toward foreign correspondents and eased legal constraints on their activity somewhat. For the native professional as well as "citizen journalists" off-line and online, however, long-standing restraints were still the norm. Interestingly, even more circumscribed were the online journalists working in the media organizations that have during the last 10 years become an increasingly important information provider for the Chinese people. Online journalists working for private news portals had no right to attend the games reporting directly from the arenas. They had to use secondary sources or watch television in order to get material for their own news items. Thus, much of the products of online journalism that reach most news consumers in China are in fact *edited* second-hand material purchased from approved state media organizations, such as New China News Agency and the *People's Daily*. Despite these constraints, since the

1994 advent of a commercial Internet to China, the interactions among various news aggregators in the online world have contributed to increased levels of transparency and public opinion in the country that has been called the "world's largest prison for journalists."[1]

The vocabulary of "health," "hygiene," and "pollution" is often used when the government describes what China's ideal Internet environment should be like. The former minister of information industry, Wu Jichuan, used to refer to the need for a "healthy Internet" (Zhang & Ni, 2000). The endeavor to establish a clean and healthy Internet was clearly illustrated when the Chinese authorities set up a website (ciirc.china.cn), especially designed to encourage Chinese enterprises and individuals to report any "unhealthy tendencies" on China's Internet. At the same time, to an increasing extent, the government also tried to nurture a "healthy nationalism" of a kind that focuses more on pride over the ongoing economic miracle and the winning of gold medals, rather than past grievances toward Japan and the United States. As the Chinese government advocates a healthy Internet, the phenomena of online nationalism and activities of journalism online are two of the more interesting themes to study in regard to Internet use in China, as they are quite interrelated. Both phenomena are signs of new demands an ever more educated and wired population puts on its authoritarian government. While nationalism adds pressure on Beijing about how to handle the foreign relations of China, journalism adds pressure on Beijing about how to address domestic challenges regarding equity, environmental degradation and free speech.

The Internet and Online Journalism in China

Ever since April 1994, when the People's Republic of China (PRC) joined the Internet, the number of netizens and alternative sources of information started to soar. As of July 2007, the Chinese networks that are accessing the global Internet were hosts to 163 million users.[2] During these 13 years living with the Internet, the regime has encouraged widespread use while it has also tried hard to keep abreast with, and contain, the pluralizing tendencies of Internet communications. In China rapid economic transformation and social change takes place alongside a much slower evolution in the political system and bureaucracy: the parts that are still supervising the mass media according to the logic of Leninist media control. This logic is also apparent when it comes to the development of new media types and forms of expression such as online journalism. In China, forms of expression and communication that were seen by overseas observers as the unfettered formation of public civic spaces during the infancy of the Internet, are now regularly checked and controlled by containment strategies, specific laws and regulations, and the technical means that are designed, implemented, and enforced by the agencies of the Chinese party-state.

In accordance with the purported needs of the party-state to uphold social stability, the security apparatus and police across the country periodically crack down on any online information or news content created that seeks, or is perceived to seek, to delegitimize the Party's hold on power. International media frequently report on how the police in China, in their quest for social order and control of runaway free speech in the digital realm, crack down on Internet cafés, close chat rooms dedicated to intellectual debate, and silence bloggers who are deemed to be too politically outspoken. In mainstream Western mass media and academia, this realm is walled in by what is often called "the great firewall of China."[3] This depiction makes sense as the Chinese system, according to some studies, is now the most sophisticated national Internet filtering project in the world.[4] Together with government propaganda, self-censorship among businesses and individual users, and cooperation between industry leaders and the bureaucracy, these mechanisms serve the purpose of containing Internet usage within the social and political limits set by the party-state.

Purpose and Methodology

This chapter seeks to shed light on the tendency of online news making to engender more openness in China, while political stability is still a primary concern for online reporters as much as it is for offline reporters. I do this by way of illustrating the case of nationalism in contemporary China, as online news portals are the common platform of both popular and state-led nationalism. On the one hand, editors and journalists have to respond to the demand from their audience that wants to have a freer discussion of news stories related to Chinese foreign policy. On the other hand, the government wants the news stories to carry "the correct message" and popular discussions related to these to stay within the boundaries of healthy nationalism. This overall trend of Chinese online journalism takes place in a professional context where the activities of so-called citizens' journalism have become more important. My focus is on the journalists' role as a broker of interests between journalists, news consumers and the state as represented by officials in the municipal propaganda departments in Beijing and Shanghai and leaders of the semi-official Internet Society of China. The research question concerns how journalists perform and perceive that role, especially in connection to issues related to foreign policy and nationalism.

For this study I have conducted 12 in-depth interviews with seven online journalists, three traditional journalists in Beijing and Shanghai and two representatives of the Internet Society of China, based in Beijing. Many of the questions raised in the interviews dealt with the increasing control of news and "the guiding of opinion" regarding Sino-Japanese relations, which were at a low point at the time of conducting field research between 2003 and 2006. The media organizations these informants work in have state-owned, municipal or private/semi-private ownership. *China Central Television (CCTV)* is state owned, while the Shanghai municipality owns the news portal *Dongfang*, and the Beijing municipality owns *Qianlong*. *Sina* is a multinational commercial portal with its headquarters in the Chinese capital. Through the interviews with journalists and Chinese nationalists that are vocal online, I analyze the interplay of journalism and nationalism, and the trend of serving the news customers by way of engaging in more hands-on journalism in China's online news landscape.

In recent years news making in the PRC has continued to avoid nonsanctioned views on politics, while sensationalism and tabloidization are news trends as they are both safe and profitable with increasing marketization of the Chinese media system. To date, with a few exceptions, not much research critically examining the development of online journalism in authoritarian countries like China exists (Wilson, Hamzah, & Khattab, 2003; He & Zhu, 2002; Chan, Lee, & Pan, 2006). And although course books and research on online journalism classes in China exist, this research tends to teach practical journalism and the commercial potential of new information technology. If not neglected, at least less attention is paid to the social and political contexts and impacts of electronic journalism.[5]

Nevertheless, despite the fact that legal constraints on journalism and self-censorship practices have successfully been transferred from traditional journalism to online journalism, it is clear that online news, news commentary and blogging changes China's media landscape (Lagerkvist, 2005, p. 130). Thus, agenda setting in the Chinese news industry is also undergoing change. Now a multitude of dissenting voices and news stories of a controversial kind are disseminated on China's Internet. Often regional or municipal online news networks follow up these news stories, and last, but not always, traditional media outlets like radio, TV, and newspapers may air or print controversial news pieces on the national level. This represents a counterintuitive logic for Western media scholars. Some believe that online journalism has more to do with creating a lively and interactive relationship with the audience, than engaging traditional media functions like agenda setting or advocacy journalism (Harper, 2003).

The Professional Identity of Online Journalists

Traditional journalists often regard online journalism as not-so-serious journalism. On the other hand, there is also an argument that the incessant move of convergence of media modalities and the advent of new production paradigms for news making leading to a very specific kind of journalism that sets online journalism apart from previous genres of journalism (Deuze, Neuberger, & Paulussen, 2004). In the West it is sometimes said that anyone can be a journalist online, hence further blurring the distinction between reliable, objective information offered by a professional and the subjective and sensationalist by the blogging citizen journalist. That logic does not hold much water in an authoritarian country like China, however, where all media, new or traditional are tightly controlled for any deviance from official policy. Not anybody in Chinese cyberspace can call themselves journalists or publish information without having the necessary education and permits. There are those that do, but they are found, fined and their websites closed and quite often they move to an overseas host server.[6]

The above-mentioned general attitudes attributed among many traditional journalists worldwide are strong also in China, not least because it has been illegal for private news portals to produce their own original content (Chan, Lee, & Pan, 2006, p. 940). They may have an editing process but the content of their news sites is often bought from state-owned media organizations. Nonetheless, they are important as news aggregators. It is apparent that the new online media environment has a strong impact, not just upon agenda setting in Chinese society, but also upon the way both traditional and online reporters work today (*CCTV* journalist, interview, January 2004).

Interestingly, to this general observation there are nuanced distinctions made by Chinese online news editors in state-owned or municipal media organizations that need to be taken into account. Taking the comparison between the print media journalist and the online journalist further to who may ultimately call themselves online journalists, editors of municipally owned *Qianlong* in Beijing and *Dongfang* in Shanghai argue that privately owned news portal *Sina*'s news staff are not "real online journalists" because they were just copying and pasting original material from others on their website (*Qianlong* online journalist, interview, 2003; *Dongfang* senior online journalist, interview, 2005).

Forms of Government Control over Online News Production

The Internet is already a platform for less constrained public opinion in China compared to the decades before the Internet. Nevertheless, the expressions that popular opinion may take in mainstream traditional media are politically path-dependent in contemporary China (Lagerkvist, 2006, p. 185). A case in point is investigative reporting that can tackle single-issue social problems only as long as their structural and political roots are not highlighted. Although critical public opinion can at times be expressed through journalism of an investigative kind and amateur citizen journalism on the Internet, it mostly stays within the boundaries set by the party-state and its propaganda organs (He, 2004). This is due to the fact that media laws in China are much clearer for online journalists than traditional reporters. There is no legal provision for independence-seeking journalism practitioners of the degree necessary to facilitate liberalization of the media (Sun, 2001, p. 87).[7] The reason is that comprehensive media laws would make the rights and obligations of both journalists and media controllers clear, and informal pressures by the party-state would then be illegal, making news control more difficult.

Today, there are two principal regulations that target the online news industry in China. The first is "Provisional Regulations for the Administration of Engagement by Internet Sites in the Business of News Publication," issued by the Press Office of the State Council and the Ministry of Information Industry in November 2000 (Baker & McKenzie Ltd., 2001, pp. 362–366). Article 5 of this law states that only websites established by press units within the state-owned media system "may engage in news publication business." Article 13 explicates in nine items what kinds of content the authorized websites may not include. Items (2), (3), (5), and (6) in article 13 are the "no-go areas" that the state agencies have most vehemently enforced. They concern, respectively, content that "discloses state secrets," "harms the reputation and interests of the state," "propagates feudal superstitions," and "disturbs social order." The second principal regulations are the "Provisions on the Administration of Internet News and Information Services," issued by the Press Office of the State Council and the Ministry of Information Industry in September 2005. This law reflects two major changes in the Chinese online news arena since the first law was issued. First, it shows both a nuanced understanding of what may constitute a news aggregator and news producer in the age of electronic media, as even bulletin board systems are required to adhere to what kinds of information they publish on a list. Thus, if the first law was more of a flexible provision able to accommodate everything and nothing, this law is more specific and detailed about who may produce news online and how.

Qianlong's news managers in Beijing do not say they serve the people or represent the views of the people by way of engaging in any kind of investigative journalism as ventured by daring editors and journalists of the Guangzhou newspaper *Nanfang Zhoumo* (Southern Daily). However, according to *Qianlong's* editors, you do not have to engage in a sensitive story on corruption to experience problems. The way the Chinese media system works still frustrates *Qianlong's* news reporters who sometimes do not see a finished story go to the printer:

> Generally speaking, when our staff of reporters is deciding on leads and issues, their ways of expressing this are different from how traditional media do it. We are also listening to some responses. When some of our reporters write articles perhaps they don't grasp all things considered when writing it up, or don't manage to express how they grasped something, then we must decide if it is publishable. We have plenty of these kinds of issues here. (*Qianlong* editor, interview, 2003).

Despite of the difficulties in having to side-step sensitive stories and issues among Chinese journalists, this particular reporter described great joy in seeing a new technology take root in the media industry, which he was sure, would lead to massive change in forms of expression. Though focus has shifted from ideology to the subtle managing of the public's attention, there is still a very perceptible government view of current affairs and politics that goes into news making. At a time when legitimacy to rule is now more important than positioning on a left-right scale, nationalistic language and pride over China's increased status in the world occupy much space in both offline and online news making. Nonetheless, online journalism and the popular commenting that is a special feature built into the genre has a potential to raise alternative awareness and even alternative mobilization, as is evident in expressions of popular national-ism. Petition campaigns, for instance—traditionally an important way for the Chinese people to voice their discontent, rally support for a cause, or attract the attention of policy makers—have moved to the realm of an online public sphere (Xu, 2007, p. 77). The petitioning tradition is aligned with a long history of advocacy journalism in Chinese culture (De Burgh, 2003, p. 34). Trained journalists or intellectuals, such as legal scholars, often take the lead in making these public movements visible and influential in the media landscape of the PRC. One obvious example of this was the case of the killing of the student Sun Zhigang, who was beaten to death in prison on the assumption he was a vagrant beggar, and how the case prompted a debate among legal scholars being activated by online public opinion (Liebman, 2005).

For the online journalist, there are considerable risks involved since a movement against governmental authority cannot normally be reported, but the political risk can be avoided if there is no direct criticism of the government, and the focus instead is on the social aspects of the human suffering. In sensitive cases, however, members of an editorial board may be unwilling to go through with publishing a particular article (Wang, 2006). One classic Chinese way to avoid political risk in order to deflect the wrath of the authorities is to use irony and subtle euphemism to disguise direct criticism. This is especially something that is visible among professional journalists that step down to engage in more independent journalism in the blogging format. As one of China's most famous journalist bloggers, Wang Xiaofeng, demonstratively wrote: "[I] don't associate with any (particular) ideas, this is just for fun." Thus, overtly, there are no politics of commitment involved on the part of these intellectual journalists that blur the line between the professional and private spheres, between social critique and sheer irony. Their disengagement from social and political realities, however, is not the most interesting phenomena. *How* China's ironic journalist bloggers, in turn, become interpreted by the public, how their ideas are acted upon, and the impact of their comments and writing are, I argue, more important. This is because these establishment journalists-cum-independent bloggers do matter as news analysts, even though they cannot be considered to be news producers or news aggregators (Yu, 2007).

Nationalism as a Great Selling Proposition

Ever since the nationalistic turn in the beginning of the 1990s, nationalism has been an economic boon for both publishing and news making (Kristof, 2001). It is, however, obvious how even the popular expressions of nationalism are subject to the needs of China's politicians. When I interviewed one of *Qianlong*'s chat room editors in 2003, he argued that he and his colleagues started to feel increasing pressure from the government to limit anti-Japanese tirades (interview, 2003). When in 2007 I interviewed one of China's most active and influential online nationalists over the last decade, he elaborated on the shift from tacit support to limits and control on the part of the central government. He stated that his organization, *China Eagle*, no longer can publish as outspoken or aggressive articles as before, as the authorities sometimes call and complain about something published as a news item on their website (interview, May 2007).

All the same, it is still much easier to publish passionate stories about national weakness as a consequence of the past errors of foreigners and potential containment of China's incipient greatness in the 21st century by powerful outsiders, than social protest in the countryside or occasional calls for democracy. Even if the government continues to monitor and suppress political opinions on China's networks, there continues to be a significant increase of pluralization of news sources and voices wanting to discuss the social and political implications of certain trends that are revealed online. These critical online discussions often deliberate on structural social ills such as officials' corruption, growing inequality between social strata, or on China's foreign policy vis-à-vis Japan, the United States, or Taiwan. Such deliberations on structural problems are rarely seen in traditional media formats, such as "letters to the editor" sections. Therefore, to an increasing extent, traditional journalists use these online discussions and deliberations when they search for news stories and valuable sources (*CCTV* journalist, interview, March 2007).

Both the limits to, and the uniqueness of, the interaction between online news editors and journalists were illustrated during the volatile anti-Japanese events and street demonstrations during 2003, 2004, and 2005. Online protesting and nationalistic diatribes against Japan and Japanese people were very prevalent in Chinese chat rooms such as *Qianlong*'s "Jinghua luntan" and the bulletin boards of *Dongfang* at the time. The policies of these two municipal news portals differed, however, as *Dongfang* editors decided to publish news stories on the demonstrations and even set up a special forum to discuss the unfolding of the protests. *Qianlong*, situated in the capital Beijing, however, took a more censorial attitude and reported less from the street protests (*Dongfang* editor, interview, May 2005).

It has been argued that nationalism is commercially viable in China. This notwithstanding, the interplay of nationalism and commercialization is also contradictory. This becomes evident through the interplay of different vested interests in the online news industry. Certain units within China's party-state political system are on the one hand intent to become a responsible global player, while other units are keen to reassure China's citizens they always protect the national interest. In this contradictory and complicated media landscape, private media organizations like *Sina* and Sohu must cater to both news consumers' demands and adhere to strict new media laws; state-owned news portals such as *Renmin Wang* (People's Net), and municipal news portals *Qianlong* and *Dongfang* must work to maintain social stability, not overly fanning nationalist flames, while they also need to attract online readers in a fiercely competitive market.

As a result, officials in charge of propaganda and news control in the State's General Administration of Press and Publication and the Communist Party's Central Propaganda Department have gradually come to understand this and they play a more subtle game than before. In the de-ideologized climate of contemporary China, nationalistic news items were until very recently what satisfied almost every interest group: the information-craving public, the legitimacy-seeking guardians of state power and the profit-seeking editors in old and new media organizations. While checks on online news making were put in place in the late 1990s, it was only since the outburst of anti-Japanese demonstrations during the spring of 2005 that the government has tried to control more extreme elements of online nationalism often expressed in the emerging arena of citizens' participatory online journalism in the blogosphere and on nationalist websites.

In general, the online deliberations, debates, and sometimes intense clashes of opinion, in their least-censored form, occur during time periods in between sensitive dates or important national events such as the annual meetings of the National People's Congress, the Communist Party congress held every fifth year, June 4, which is the date the Tiananmen Massacre in 1989 took place in Central Beijing, or the death of leaders of a politically progressive inclination.

Before and during such events, online journalistic activity is more restricted as are postings to bulletin board systems (BBS) and chat rooms.

Scholars have argued that the Chinese government is moving away from the propaganda model, and that the Party's dominant role is increasingly difficult to uphold (Latham, 2000, p. 654; Lee, 1990, p. 3) or that as the Chinese propaganda machine has come under increasing attack from the forces of commercialization, it fails to uphold its old idea of disseminating "thought work" to the population (Lynch, 1999, p. 2). Although this seems to be the case for online popular nationalism as expressed on websites and bloggers columns (Xu, 2007, p. 156), they too have increasingly become supervised by propaganda units in the Chinese media system that monitors any transgressions of official foreign policy (*China Eagle* leader, interview, May 2007). The reason is that their nationalist activities and Internet forums have become so influential that news websites started to give them a lot of attention. There is, however, something deeply misleading in dismissing the role of propaganda in China (Lagerkvist, 2006, p. 158; Brady, 2007), as even passionate popular nationalism rather easily are swept away and a more "healthy" official version of responsible nationalism takes its place. Despite the enormous growth of the Internet in China most citizens and journalists conform to the hegemonic position and line of the Communist Party when called for. The term "public supervision," which is frequently used now by the official media, represents an attempt by the Chinese government to manipulate and manage public opinion in a softer and subtler way than before (Chan, 2002, p. 50).

Both the elderly, who learned how to see and read through old-style Maoist propaganda, and the young in search of firm identities, are subtly persuaded by new smart information produced by the spin doctors of the Communist Party as China's leaders are chartering their political course. I have termed the Chinese party-state's propaganda strategy on the Internet, *ideotainment,* which consists of two forms; hardly noticeable covert persuasion, or overt persuasion demonstrating the correctness of government policy (Lagerkvist, 2008). Ideotainment tactics are often covert nowadays with political messages and symbols often blended with seemingly straightforward news, nationalist sentiments, and formats of popular culture. Overt ideotainment using the old-style persuasion techniques as employed in the campaign against the Falun Gong between 1999 and 2001 (Toong, 2005, p. 207) still exist in times of political tension. Ideotainment as a term captures a *process* of subtle propaganda in the information age, in which both authoritarian and democratic countries act to "bend" information, and consequently public opinion, according to the will of governments. In today's China, we see this also taking place in the delegation of routine surveillance and self-censorship among Internet service, content providers, and even among so-called ironic bloggers, as is clear from the following quotation:

> Actually, we work very, very closely with the government, the propaganda department, or the Ministry of Information Industry. We follow the guidelines and have to be working partners with the government. It is not a choice to work with them, but on the other hand, it is a choice. . . . we are not allowed to have our own correspondents; we are dependent on the government for news. It is a good way—saves cost also. (Manager of the private news portal *Sina*, interview, 2003)

That the social control practices in the work units of state-owned companies no longer exist, or lack the importance they once had, does not mean that social control practices in the new market economy are nonexistent. How officials intend to strike a balance is evident from the above quotation that illustrates these evolving practices of delegated censorship among media organizations in China. Surveillance practices of dissident, thought, are now partly being transferred to media companies such as Nasdaq-listed Chinese news portals *Sina* and Sohu who are working in tandem with authorities to hunt down any information that is deemed immoral, breaks relevant Chinese laws, or that uses overly critical language of the central government in Beijing. Not only domestic companies are cooperating with the authorities in this quest. The American Internet giant Yahoo has been implicated in delivering personal information about Chinese users of Yahoo's email accounts.[8] Microsoft and Google are regularly accused for kow-towing to the Chinese authorities in exchange for market access.

As a matter of fact, propaganda officials and online news editors of *Qianlong* in Beijing and *Dongfang* in Shanghai participate in a mutual learning process, in which the former get to know the commercial logic behind news production, which pushes old-style propaganda into the background. The latter still live with constraints on reporting sensitive news, as well as occasional demands to produce softer propaganda more in touch with ordinary people's lives. Thus, online journalists and editors are shouldering the role of "chief negotiator" between their two masters: the Communist Party and the public. This peculiar double role means that they broker new space for public opinion, while at the same time they are also charged with the responsibility task of building legitimacy for the policies, programs, and ideology of the party-state. There is a certain strand of Confucian thought built into this responsibility system, according to which journalists must assume not just the negotiators role but also the task of keeping society stable (Lagerkvist, 2008). The online news portals, both state-owned and private, serve as an intermediary negotiating the interests of the party-state, the information-seeking citizenry, and naturally also their own interests in the process of making news online.

Contrary to much conventional wisdom, however, my interviews revealed that municipal news portals are under less pressure and are experiencing greater maneuvering space than private and semi-private news portals such as *Sina*, *Sohu*, and *Netease* (*Dongfang* news editor, interview, May 2005 and May 2006; *Shanghai Online* managing director, interview, October 2003; Internet news editor of Xinhua News Agency, interview, May 2006). While the latter dared not to publish any stories on the nationalist demonstrations against Japan in 2005 (*Sina* senior news manager, interview, May 2005), the municipal news portals did, especially the Shanghai-based *Dongfang*. The private news portals have to attract news readers by way of breaking earlier social taboos on nudity instead, which my informants argued rendered state and municipal news portals more authoritative, whereas the private news portals are seen as sensationalist (*CCTV* journalist, interview, January 2004). Thus, although the private media elite belongs to an emerging bourgeoisie middle-class stratum, which, according to models of civil society, should represent the voices and interests of critical post-materialist citizens, they do less than the official news portals to push the boundaries of social and political news making online.

Conclusion

My findings show that the media elite in state-owned or locally owned news portals, reformist e-government officials, and policy makers and propaganda officials influence each other in ways that tend to cede concessions of social freedom and space for free speech in exchange for understanding of efforts needed to maintain the stability that is so cherished by the Communist Party (*Qianlong* journalist, interview, October 2003 and January 2004). The negotiated result is a controlled online news industry with journalists shouldering a role of broker between various state and popular interests in society: a so-called healthy Internet where all kinds of smart propaganda blend with popular culture to create a mix of the "correct" political views and "healthy nationalism." By contrast, the managers of the privately owned news portal *Sina* tread very carefully, taking no risks in possible affronts to the authorities.

The proposition of a collaborative framework of cooperation and negotiation for the online news industry is consistent with the way the online news industry works in order to control to what extent news stories related to sensitive foreign policy issues are allowed to be published. The trade-off between commercial and political logic reveals itself both in the quantity of nonpropaganda-related material and in the content and scope of ideology in the party and policies of government. The party-state wishes to save and modernize the "commandist" media system by domesticating some nonroutine measures associated with the market economy.[9] Through such collaborative efforts, the overall framework of the commandist system is maintained and reproduced, and along the way, some forms of "nonroutine measures" are incorporated, which is helpful for expansion of the public sphere. Partly with the help of feedback from online news media, the propaganda system (*xuanchuan xitong*) learns about social pluralization and how to design a "guidance of opinions" (*yulun daoxiang*) which is becoming increasingly effective; given this, further liberalization of the Chinese media sector will be more the result of domestic market competition than globalization.

The regulation "Provisions on the Administration of Internet News and Information Services," issued in 2005 showed a more liberal stance vis-à-vis media organizations that were previously only able to purchase news items from official sources. According to this regulation, they could produce their own news items, provided they showed that their editors were educated journalists with some experience. This more liberal policy of the authorities can be attributed to the fact that the government's strategy to control online news making has become more effective. It isn't only the state-owned or semi-private media organizations in the new media sector that acquiesce to the Communist Party line on media control. There is indeed a broad delegation of routine surveillance and self-censorship among Internet service and content providers (Lagerkvist, 2006, p. 167).[10] On the issue of individual self-censorship, there is evidence that the overwhelming majority of individual Chinese netizens and those responsible for intellectual chat forums engage in this practice, too (Zhou, 2006, p. 179). The fact that the Chinese government face no imminent threat from an online public sphere may be the reason for the more liberal tendency of allowing new media organizations to engage in journalistic activities.

In order to gain more knowledge about the continued trade-offs between journalistic freedom and media control practices in an increasingly commercialized context, interesting avenues for future research on online journalism in China include gaining deeper insights into, on the one hand, the phenomenon self-censorship in China's new media organizations, and on the other, an examination of how their activities and innovative formats simultaneously lead to more knowledge, debate and popular demands for political accountability. Particularly promising would be newsroom studies and focus group interviews with targeted groups of Internet news consumers and bloggers who regularly interact with online journalists.

NOTES

1. See the reports of the Committee to Protect Journalists on China and journalism at www.cpj.org and Bridget Johnson's summary at About.com: http://journalism.about.com/od/pressfreedoms/ig/China--Prison-for-Journalists

2. See China Network Information Center (CNNIC): www.cnnic.com.cn

3. See CPJ (2002), "The Great Firewall: China Faces the Internet": http://www.cpj.org/Briefings/2001/China_jan01/China_jan01.html

4. See OpenNet initiative (2006), "Internet Filtering in China in 2004-2005: A Country Study": http://www.opennetinitiative.net/studies/china

5. See for example Zhong (2002).

6. One case is Yulunjiandu.com, a website established by a private individual whose calling is to expose official corruption and publish all available evidence online.

7. It should be noted though, that according to article 35 in China's constitution, all citizens of the PRC enjoy freedom of speech and freedom of the press. According to article 41, PRC citizens have the right to criticize any state organ or functionary. See the Constitution of the People's Republic of China: http://www.usconstitution.net/china.html

8. See "Yahoo! Implicated in Third Cyberdissident Trial, US Company's Collaboration with Chinese Courts Highlighted in Jiang Lijun Case": http://www.rsf.org/article.php3?id_article=17180

9. For a description of the term "commandist system" describing the Chinese media system, see Lee (1990, p. 14).

10. See also the joint statements between the IT-businesses and the Chinese government: Zhongguo wangluo meiti de shehui zeren, Beijing xuanyan zai Jing qianzhu [The Beijing Declaration on the Social Responsibility of China's Internet Media is signed in Beijing]: http://www.thmz.com/html/col82/2003chinalt/1003.htm

CHAPTER NINE

Do Online Journalists Belong in the Newsroom? A Belgian Case of Convergence

Vinciane Colson & François Heinderyckx

Colson and Heinderyckx report preliminary findings from research with Belgian media which continued at the time of this publication. Although they report the results of ethnographic investigation of an online newsroom—as with the rest of chapters in this section, their focus is on convergence. The study provides an up-to-the minute analysis of specific convergence processes underway (and the organizational stresses they foment), complementing the overview of convergence research provided by Singer in the next chapter. It is a timely, and cautionary, analysis, coming at a time when many media institutions are struggling to find the right strategy for integrated production. The disconnect they find between management decree and news production practice echoes a common theme of studies presented here.

EDITORS' NOTE

As soon as the use and adoption of the Internet in households and businesses reached a sizable level, all media organizations felt compelled to establish a significant presence on the Internet. After clumsy attempts to use these new platforms to promote their outlets and make some of its content available online, it became clear that the World Wide Web and the surrounding information and communication technologies would lead to innovative ways to create and circulate content.

Within news organizations, staff had to be redeployed or even hired to take on the new tasks associated with developing a presence on the Web and with feeding the online outlets with content. At first, these tasks were thought to be of a mostly technical nature. Files were to be imported, converted, sorted, and arranged. As the interfaces grew in complexity and the expectations of the users grew in sophistication, the skills required to maintain an online presence outgrew that of mere computer programming and graphic design. Online content called for specific writing and formatting. A parallel news flow was emerging alongside that of the main newsroom, with its own processes of selection, sometimes its own editorial policy, and its own identity. Although this must all remain compatible and somewhat synchronized with the main newsroom, staff assigned to the management of the online presence was unquestionably dealing with a number of tasks and responsibilities of a near-journalistic nature. In the late 1990s, a new trade was emerging in news organizations, that of online journalists.

Meanwhile, journalists were asked to adopt and use more and more digital technologies in their work. As other chapters in this book demonstrate, they became gradually familiar and skilled at using tools and technologies that, combined with a reshuffling of the workflows, would eventually facilitate the migration of their work to the various digital distribution channels, starting with the Internet.

As a result, a technology-centred work force gradually took on tasks of an editorial and journalistic nature, and journalists were expected to become even more technologically literate—capable of covering a wide span of the multimedia outlets that their organization is to feed. A good number of self-proclaimed web or online journalists considered themselves a "special breed" of journalists (Deuze & Dimoudi, 2002, p. 96).

This concern illustrates a challenging set of issues at the level of human relations, namely those between print journalists and their online counterparts. It is the case that the technological barriers to deploying a fancy website are often simpler and more predictable than the reactions of staff forced into changing their long-standing work habits as well as into redefining their identity.

Despite being in existence for up to a decade, professionals preparing and providing online content (hereafter "online journalists") are still often not seen as fully-fledged journalists. Traditional journalists tend to be reluctant to consider their online counterparts as peers (Singer, 1998, p. 7). Some of the reluctance might come from tasks performed within online journalism, which is still "first and foremost a desktop job" (Deuze & Paulussen, 2002, p. 241). In the case of newspapers, the online journalists are heavily dependent on the print version of the newspaper and spend more time rewriting and reformatting content prepared for the printed version than doing actual reporting.

In Europe, the status of online journalists became an issue only recently. In November 2006, after months of talks between trade unions and the management, the online journalists at U.K. newspaper the *Guardian* were finally granted the same salary as the other journalists. The agreement marks a major step toward the consideration of online journalists as professional journalists in a context where an increasing number of newspapers opt for converged newsrooms where online teams are relocated within or just next to the newsroom, eventually merging their web and paper newsrooms into a single entity. Recent examples include the *Daily Telegraph* or the *Financial Times* in the United Kingdom and the *Volkskrant* in the Netherlands.

In Belgium, questions of the upgrading of online journalists' status and of editorial convergence are a leading concern of the media industry. This chapter describes an investigation of the attempt by Belgium's newspaper *La Libre Belgique* to take advantage of a relocation to physically integrate the website team into the newsroom. The question is: Does this cohabitation mean that online journalists have achieved parity with their traditional peers? This ethnographic study, based on observations of the work in the newsrooms and interviews of journalists, aims to identify the changes brought about by this amalgamation. It tries to analyse the relationships between online and paper journalists as regards issues of cooperation, aspirations and mutual representations, in a converged newsroom.

Media Convergence

These reorganizations occur within the broader phenomenon known as media convergence, which essentially enables the newsroom to provide information and content in a variety of formats to feed the different media forms and distribution channels and news outlets. According to Zollman (2001), cooperation between two newsrooms could be considered a form of convergence. For example, he argues that a fully integrated newsroom should not happen until "the audiences and revenues of traditional and new media services are more evenly balanced, the interactive media are making solid profits, electronic production systems are able to handle multiple digital media assets across platforms with ease, and when business models have become fully established" (Duhe, Mortimer, & Chow, 2004, p. 86). Jenkins (2001, p. 93) defines media convergence as the combination of at least five convergences: technological (digitization), economic (cost savings by horizontal integration), social (adoption of the multimedia channels by the users), cultural (new forms of creativity and transmedia storytelling), and global (international flows of content). We will later argue based on our field study that media convergence requires even more than these five cumulative features to be fully achieved.

A number of studies have considered the consequences of editorial convergence on newspaper content. In his ethnographic study of three online newsrooms, Boczkowski (2004b) shows how variations in organizational structures, collaborations between print and online journalists, work practices and representations of the users are related to the different ways in which members of the newsroom appropriate interactivity and multimedia. Singer (2004a, p. 838) examined the consequences of convergence on the print journalists, and in particular their resocialization in a converged newsroom. But few studies have contemplated the ensuing changes in the journalist's daily work and the relations between the two types of journalists. In 2000, Huxford and Duda explored the question and concluded that these relations could be defined as "a collision between cultures." Based on their ethnographic study of three U.S. newspapers (the *New York Times,* the *Philadelphia Inquirer,* and the *Morning Call* of Allentown, PA), they identified many differences between print and online journalists. Print journalists were found to be "usually middle-aged, graduates of journalism schools, have been working at their newspapers for extensive periods of time," defining their own work in terms of mediation or gatekeeping. Online journalists, on the other hand, are "usually younger, graduates of many different schools, tend to change their jobs every few years" (Meyers, 2002, p. 25). According to Huxford and Duda (2000), these discrepancies are conditions likely to produce a clash of cultures.

Our research considers this "collision between cultures" in the context of an in-depth reorganization at Belgium's daily *La Libre Belgique.*

Case Study: *La Libre Belgique*

Based on observations on the development of the newspaper and its website, this study aims to determine the extent to which online journalists belong in the newsroom.

The prominent and influential Belgian French-speaking daily, *La Libre Belgique,* owned by the Belgian corporation IPM, provided an opportunity for ethnographic observation when it recently took advantage of a complete relocation to physically integrate the website team into the main newsroom, alongside journalists. By using "methods which capture the social meanings and ordinary activities" (Brewer, 2000, p. 10), this research approach appeared the most appropriate way to achieve a qualitative insight into the journalists' daily work and how the attempt to graft online journalists into the newsroom was organized and perceived. As ethnography calls for the combination of multiple techniques of data collection (Cassell & Symon, 2004, p. 313), this case study had three parts. Firstly, Colson conducted an exploratory interview of *La Libre Belgique*'s editor in chief, which allowed us to prepare the second stage of data collection: nonparticipant observation of work practices which was conducted over three days in 2007.[1] We especially paid attention to exchanges and interactions between online and print journalists. Finally, seven open-ended interviews were conducted with actors from the field.[2] Based on the collected material, we have analysed the situation with respect to issues of cooperation, aspirations, mutual representations, work flow, and social practices.

Created after World War I, *La Libre Belgique* faced challenging times with a circulation of less than 50,000 copies in the second quarter of 2007 (to be compared with over 80,000 copies of the popular daily *La Dernière Heure*, owned by the same group, and over 90,000 copies for the competing French-speaking quality paper *Le Soir*).[3] As with so many newspapers, *La Libre Belgique* is actively seeking new strategies to enlarge its audience and consolidate its revenues.

The website of *La Libre Belgique* (www.lalibre.be), after a late start (2001 while competing *Le Soir* started its website in 1996), is now quite successful and in constant growth with over 460,000 unique visitors for a total of nearly five million page requests in July 2007. On average, about 22,000 regular users visited the website every day in July 2007, which is significantly less than the 39,000 of *La Dernière Heure* (www.DHnet.be), designed and managed by the same team.

Erratic Convergence

To better understand the development of the collaboration between print and online journalists in *La Libre Belgique,* and to better analyze our observations, we must consider the successive relocations of the website team within the premises of the newspaper offices. After years of negotiations with journalists on issues related to intellectual property and royalties, *La Libre Belgique* managers created their website in 2001. The website was welcomed by print journalists who saw it "as an extension of the paper, a good manner of attracting new readers to the newspaper." That first period seems to be considered as an observation period by *La Libre Belgique* workers.

A year later, in an attempt to rationalize resources and enhance the sharing of experience, *La Libre Belgique*'s online journalists were moved two floors lower to join the website staff of the popular *La Dernière Heure*. However, print journalists of *La Libre Belgique* were very critical of the work of their colleagues at *La Dernière Heure* and felt detached from their own website. Furthermore, the absence of physical proximity and the resulting lack of personal interaction increased this feeling. The short distance between the newsroom and the online team was enough to limit dramatically exchanges to a few remarks or complaints exchanged via email and telephone.

In January 2007, *La Libre Belgique* and *La Dernière Heure* moved to new premises. Journalists of both newspapers were relocated in a large open space which, although preserving a clear separation between both editorial teams, looks almost like one big newsroom. In order to mark a clear demarcation between the workspaces of journalists of the two newspapers, and in an attempt to refit the online team within the newsroom, thus implicitly endorsing that they were journalists among other journalists, the workstations of the online journalists were located in between, effectively acting as a buffer between journalists of both newspapers. Convergence of workspaces matched the aspirations of online journalists who felt increasingly at odds with the lack of recognition of their status after several years of being immersed into administrative services, alongside sales and advertising departments and away from the newsroom. We call this period the "grafting attempt." Yet, proximity did not lead to parity, and the graft did not take. In just five months, the planned reconfiguration has evolved in unexpected ways and forced an immediate relocation of the online staff.

Despite their central location in the middle of the newsroom, online journalists were largely ignored. "We feel as if we were invisible," said one. Our interviews indicate that the older print journalists appeared somewhat fascinated by the new technologies, while the youngest ones, digital natives, educated in a world of technology, showed less interest in the web-related fuss and seemed less interested in convergence. This, if confirmed by more observations, would relativize conclusions by Singer (2004a, p. 850) that journalists with extensive experience would be reluctant to cross the bridge.

From January to May 2007, the online journalists physically cohabited in the newsroom but we saw that they did not actively participate in the newsroom's life. They were not invited to the editorial meetings, even though it could have improved the quality of the website, its coherence with the paper and the work flow inside *La Libre Belgique*. Eventually, they were invited to such a meeting once, but had to turn down the invitation because the time of the meeting was not compatible with their own work (a heavy workload of updating the websites in the morning).

After two months of cohabitation, we could perceive change. Curiosity seemed to be growing among some print journalists who occasionally reached out and made contact with the websites desk. Occasionally, some collaboration seemed to emerge (e.g., a phone call after a press conference from a journalist to alert his online peers). Some journalists frustrated by a shortage of editorial space asked if they could elaborate further on the website. However, tension was still visible when, for example, the editor in chief allowed the online team to prepublish an exclusive on the website.

After only five months of this grafting attempt, management decided to abort it and move the online staff out of the newsroom and into the graphic design unit, thus symbolically and implicitly bringing the online team back to its mere technical facet. Interestingly, no reasons or explanations were provided to the workers. The web staff was simply informed that it would be moved and the rest of the journalists simply noticed their removal. While this removal is seen as a punishment and a step backwards by online journalists, the print journalists have noted their disappearance without emotion. "One morning, a week or two ago, I don't know, I noticed that there wasn't anybody at their desk. I don't know where they are," explained one print journalist.

When asked, *La Libre Belgique* management described the decision as carefully weighed and strategic: the online journalists are to work more with the graphic team to improve the quality of the ergonomics and graphics of the website. Furthermore, the radio journalists from *Radio Ciel*, owned by the same publisher (IPM), will also be located in the same section of the building in what amounts to a new attempt at convergence, this time between web and radio journalists.

The cooperation and collaboration between the newsroom and the website team is now back to square one: limited contacts mostly by email. Later during the second phase of our observations (after the relocation of the online staff outside of the newsroom), we saw a print journalist go upstairs to the online newsroom, but only to complain about how his own article was featured on the website and about an erroneous hyperlink.

Convincing overworked journalists to add to their workload by writing for the website is a challenge. In order to encourage emulation, it was decided that one journalist will be given the task to act as an intermediary between the print journalists and the website content editors. Every day, he is expected to talk to journalists to determine what they can do on the website and to improve the quality of the information provided. For important stories, they will prepare special reports exclusively for the website. The goal is also to create inside the print newsroom a sense of reactivity and continuous updating so typical of online journalism. The effectiveness of this mediation remains to be seen.

Findings

In terms of mutual representations, our observations showed a pattern of "us" versus "them" representation, between online and print journalists, in spite of the physical proximity during the "grafting attempt." Even if the relationships showed encouraging signs of improvement, print journalists still do not view online journalists as peers, while the latter perceive the attitude of print journalists as haughty and disdainful.

In the course of our observations, we perceived a marked disparity of concerns and aspirations between the two groups. While print journalists worry about their articles, the veracity of information and their sources, online journalists' immediate attention is absorbed by time-consuming concerns about the website's pages settings and layout, the renewal of the top stories or the moderation of the forums. "We don't have time to verify every news wire before putting it on the website," notes the editor of *Lalibre.be*. "Our priority is to react quickly. On the Web, if some information is erroneous, it's easier to rectify." This study confirms the assumption, supported by the surveys of Deuze and Paulussen (2002, p. 243), that online journalists are less concerned with their roles as gatekeepers and agenda setters than traditional journalists.

The overall situation is largely reminiscent of the one described by Huxford and Duda in 2000: the relations between the print journalists and the newer breed of online journalists could be described as a collision between cultures (defined by Singer as "the set of shared attitudes, values, goals and practices characterizing a social or occupational group"). "For newspaper journalists, it encompasses professional values of expertise, ethics, public service, and autonomy, plus work routines that foster those values" (Singer, 2004a, p. 846).

The attitude of *La Libre Belgique's* management was also similar to that observed in their study. The decision to integrate the online staff into the newsroom was reportedly taken just a few days before the relocation, and then, the decision to move it back away from the newsroom was taken only five months later. The managers took the decisions at an institutional level, without preliminary consultation with the society of journalists[4] at the newspaper. They attempted to artificially create convergence with a lack of consideration for the opinion and sensitiveness of the workers, and with no clear long-term vision or strategy outside of mere logistical concerns. Huxford and Duda (2000) insist that in the newsrooms they surveyed, "the online innovation was almost instantly adopted at the institutional level. In all three newspapers, there is also some degree of investment in the creation of independent online editions. While journalistic institutions adopted the concept of online journalism enthusiastically, the journalists were more sceptical" (Meyers, 2002, p. 25).

The journalists did not receive training to help them adapt to the changes. "We don't mind writing for the website or recording our interviews for podcasting, we don't have the choice. But, to edit these sorts of material, we need learning and training," explained one print journalist from *La Libre Belgique*. By the time of this writing, no training had been organized for online and print journalists to help them use a sound recorder, a website, a photo, and video camera and new technologies in general.

Our research indicates that online journalists do not yet belong in the *La Libre Belgique* converged newsroom. The failed grafting attempt sheds light on the process of media convergence within that organization. Based on the five-process typology developed by Jenkins (technological, economical, social, cultural, and global), the Belgian newspaper could be seen as progressing toward achieving media convergence:

Technological convergence: In November 2006, the IPM group had adopted an entirely new software platform for content to be used not only for printed publications, but also for the full spectrum of multimedia forms, enabling and encouraging the exchange of content between print and online journalists. By doing so, *La Libre Belgique* and *Lalibre.be* reached technological convergence.

Economic convergence: From an economic point of view, the publisher IPM already maximized the convergence between its different outlets: all actors are now grouped in a single location, including a radio station (*Ciel Radio*) as of 2007. IPM has achieved horizontal integration and by doing so, has reduced its operating costs.

Social convergence can only be speculated on in the absence of specific research, but one can assume that visitors of *LaLibre.be* are not significantly different from the rest of Internet users and that they do, indeed, make their visits to the site just one of several experiences within their media diet.

Cultural convergence as "development of content across multiple channels" (Jenkins, 2001, p. 93) is more evident in *La Libre Belgique*'s newsroom. Despite a lack of enthusiasm among journalists regarding the Internet as a news medium, print journalists are increasingly aware that their work will be more and more transposed in various formats and distributed by variety of channels. They also understand that the website has a more flexible structure which allows them to overcome the rigid limitations of the newspaper's editorial space. In the course of our observations, we noticed that journalists occasionally used the website version to add details of an interview that they could not include in the paper version.

Global convergence: Observing the work of the journalists (both print and online) at *La Libre Belgique* shows clear signs of global convergence as defined by Jenkins. Journalists in the quest for information and content rely heavily on a variety of online content providers including other news media, content sharing platforms, news concentrators and news agencies (now taking full advantage of convergence). When working on an international story, journalists are keen on consulting the websites of leading international media.

Based on Jenkins' criteria, one can speculate that *La Libre Belgique* and *Lalibre.be* are on their way to media convergence. Yet, the failed attempt to graft the online staff in the newsroom leads us to believe that one crucial factor is still hindering the company in this process of convergence. We should therefore consider that a sixth dimension is required to achieve full convergence: *editorial* convergence, which implies a balanced cooperation between online and print journalists along with a reciprocal positive representation of each other's work and role. This sixth dimension of convergence is to be seen as a keystone to fully-fledged convergence, even though, in the case examined here, management appeared more concerned with technology and logistics.

Conclusion and Discussion

At each stage of development at *La Libre Belgique,* management made important decisions with hardly any discussion with the journalists. The successive removals of the website team were imposed on the journalists overnight, without any substantial explanation. Some of the barriers existing between online and print journalists may be rooted in the lack of communication by the management about its multimedia strategy, a deficit in training of print journalists, and the resulting lack of commitment by staff for the new configuration and expectations associated with the unavoidable convergence.

Clearly, online journalists do not yet belong to the newsroom at *La Libre Belgique.* But was it really the strategy of the IPM group? The editor in chief has informed journalists that they must now write for both the paper and the online versions of the paper. They have to become "multimedia journalists." He explained that he would like a newsroom producing content which comes in a variety of interoperable forms, i.e., which can be made available on the different distribution channels (website, paper, radio, webTV, mobile phone, etc.). In the future, he speculates, the website team will focus exclusively on the breaking-news feed.

Our observations have led us to establish that Jenkins' five-process convergence dimensions do not account completely for all the aspects of media convergence. In addition to technological, economic, social, cultural, and global convergences, we found that a form of "editorial convergence" must also be taken into consideration. Short of reciprocal positive representations and constructive cooperation between print and online journalists, a fully integrated newsroom cannot be achieved.

Despite the proximity provided by a shared workspace, the relationships and collaboration between the quality paper's newsroom (*La Libre Belgique*) and the popular daily's newsroom (*La Dernière Heure*) remain fraught and counter-productive (no cooperation, no common aspirations, no contact "except near the coffee machine"). Although the first attempts to develop synergies with the website failed, subsequent effort to enhance the role of the online staff to unite the different editorial teams should be further investigated. The role and feat of the new "intermediary" who will be made responsible for actively interfacing between the print newsroom and the online team will be of particular interest.

NOTES

1. This observation phase was preparatory to the larger study still in progress.

2. We conducted interviews with the president of La Libre Belgique's Society of Journalists, the editor of the website, one paper journalist, three online journalists, and again with the editor in chief.

3. Stats of newspaper circulation and online audience by CIM (www.cim.be).

4. An association gathering journalists of a particular medium with the primary aim of defending their independence from the company managerial decisions and the quality of their work. Not to be confused with trade unions or professional bodies.

PART THREE

Reinventing Journalism?

CHAPTER TEN

Ethnography of Newsroom Convergence

Jane B. Singer

Newsroom convergence is probably the most-discussed trend in journalism after online journalism itself. Attitudes and strategies range from abstract acceptance to practical resistance, and definitions of what convergence might—or should—entail vary widely. This chapter summarizes what research to date indicates about convergence, discussing the benefits of an ethnographic approach and its stress on the processes and narratives of the actors involved.

EDITORS' NOTE

Newsroom convergence offers a case study for the value of the case study: It is an ideal subject to explore through ethnographic methods that can comfortably mix and match participant observations, in-depth interviews, document analysis, and subject surveys (Lindlof & Taylor, 2002; Wimmer & Dominick, 2003). This chapter examines why the topic so readily lends itself to an ethnographic approach, illustrates how researchers are using it to understand this profound change in newsroom culture, then looks more closely at one series of ethnographic studies. It concludes with a brief consideration of the ongoing utility of ethnography as newsroom products and practices continue to evolve.

"Newsroom convergence" is a problematic term, both because it has fallen out of favor in recent years and because it includes many different activities and workplace structures. It is used here to describe a change from single-platform journalism—creation of content for a newspaper or a television news program, for example—to cross-platform journalism involving more than one medium. In reality, news organizations may lie anywhere on a continuum from cross-promoting stories to planning and producing coverage keyed to the strengths of each medium (Dailey, Demo, & Spillman, 2005). Virtually all news organizations in the developed world now provide content for a traditional media outlet and an affiliated website. A relatively small subset, the focus of much convergence research in the United States, are asking journalists to generate content for three media—typically, television, a newspaper, and a website. Organizations that own both print and broadcast or cable outlets in a market are behind many of these efforts; others involve partnerships between unaffiliated newsrooms. Either way, journalists in the newly converged newsrooms previously competed for stories rather than cooperating to produce them.

More broadly, however, "convergence" is occurring even in newsrooms with no external partners at all. Given the rapid development of broadband and video technology, the addition of a website to the news mix makes nearly all journalists cross-platform storytellers. Television journalists must write text versions of stories they previously told with sounds and images. Print journalists must gather, edit, and incorporate audio and video material. Whether or not these television journalists contribute to a newspaper, or print journalists to a television news program, they must learn to communicate effectively using a more extended vocabulary of media technologies than before.

Such changes present many challenges for journalists—and interesting questions for scholars. The next section outlines why the ethnographic approach offers an opportune way to probe for answers by highlighting key attributes of this flexible and multifaceted method.

A Good Fit

Qualitative ethnographic methods, drawing on participant observations and in-depth interviews of working journalists, have been widely used to explore newsroom convergence. Surveys also have been common, and while some such studies are clearly not ethnographic, others are; quantitative data collected during ethnographic observation are included in the fieldwork concept that is central to the approach (Delamont, 2004).

This section looks at several hallmarks of ethnography and considers their fit with scholars' desire to understand how convergence affects journalists' practices and perceptions, in turn shaping an emerging identity of multimedia journalism (Deuze, 2004).

Ethnographers study the meanings of behavior, language, and interactions among members of a culture-sharing group (Creswell, 1998). The centrality of culture is the heart of the ethnographic approach, particularly as applied to groups facing restructuration and a loss of traditions that may erode earlier certainties (Willis & Trondman, 2000).

This cultural focus of ethnography is crucial to the study of how journalists perceive and adapt to changes in the way they make news. Media managers and practitioners have declared their occupational culture the hardest thing to change and "cultural resistance" the biggest hurdle to overcome (Thelen, 2002, p. 16). Convergence scholars have almost universally been concerned with how journalists are negotiating this challenging cultural transition (Jenkins, 2006).

Technological change affects how journalists do their work; the content they produce; the structure of their work environments; and their relationships with sources, competitors, the public and one another (Pavlik, 2000). Although ethnographers largely neglected technological dimensions of newsroom culture in the Internet's early years (Boczkowski, 2004b), recent case studies have explored how changing production modes and methods affect the quality of news reporting (Huang et al., 2004) and journalists' perceptions of their public service role (Singer, 2006a). Silcock and Keith (2006) focused on language issues related to efforts to converge print and broadcast news operations.

Ethnographic methods put the researcher in the middle of the topic under study: The researcher goes to the data rather than the other way around (Wimmer & Dominick, 2003). The primary aim is "to describe what happens in the setting, how the people involved see their own actions and those of others, and the contexts in which the action takes place" (Hammersley & Atkinson, 1995, p. 6).

Fieldwork has been central to the study of newsroom convergence, with researchers observing workflows and talking with journalists within their work environments. The News Center in Tampa, Florida—where Media General invested $40 million in the late 1990s to build a "temple of convergence" (Colon, 2000) housing print, broadcast, and online news operations—has been a particular place of academic pilgrimage. At least half a dozen ethnographies published in refereed journals have focused on convergence in Tampa either alone or in combination with a small number of other news organizations.

Research within converged newsrooms has produced a deep understanding of what the change has meant to those engaged with it. For instance, Boczkowski's close observations and extensive interviews within three news organizations led to his insight that convergence is "a contingent process in which actors may follow diverging paths as a result of various combinations of technological, local, and environmental factors" (2004b, p. 210). Lawson-Borders (2003) amassed more than 36 hours of taped in-depth interviews during her fieldwork in converged newsrooms, enabling her to develop a set of best practices for news organizations seeking to assess and implement the change.

Ethnographic studies are characterized by a need to remain open to elements that cannot be codified at the time of the study. This openness is necessary if researchers are to discover how the people they are studying understand and represent the world (Baszanger & Dodier, 2004).

Scholars rarely write about their own thought processes, so specific examples of ways in which this exploratory mindset has informed the study of newsroom convergence are elusive. However, the desire to convey subjective understanding of convergence is often referenced explicitly in abstracts and other authorial framing devices. Dupagne and Garrison (2006) frame their study as an attempt "to understand the meaning of this media convergence experiment, the changes in the newsroom culture, and the type of job skills necessary in a convergent newsroom" (p. 237). Silcock and Keith (2006) say their goal is to determine how convergence is defined by the journalists involved in it; Singer (2006a) similarly describes a desire to understand journalists' perceptions.

A number of convergence ethnographies also recommend quantitative follow-ups, indicating their initial work has sought to identify concepts to which more narrowly formulated approaches might subsequently be applied. Dupagne and Garrison (2006) call for a survey of journalists in converged newsrooms "to determine the perceived importance of traditional and convergence job skills" (p. 252). Lawson-Borders (2003) and Singer (2004a, 2006a) both recommend longitudinal work involving qualitative and quantitative methods to track the evolution of convergence.

Ethnography typically involves in-depth investigation of a small number of cases, perhaps just a single case rather than trying to represent general trends (Atkinson & Hammersley, 1994). The purpose is to represent not the world but the case, a bounded and integrated system with identifiable patterns of behavior (Stake, 2005).

News organizations nicely fit this definition, and most of the published ethnographic work related to newsroom convergence draws on one to four cases. Changes in journalists' behavior patterns have been of particular interest. Examples of single-case ethnographic studies include Cottle and Ashton's (1999) exploration of early moves toward convergence at the BBC; Dupagne and Garrison's (2006) consideration of the *Tampa Tribune*; and Meier's (2007) work at an Austrian news agency. Other convergence studies have looked at two (Silcock & Keith, 2006), three (Lawson-Borders, 2003; Boczkowski, 2004a, 2004b), or four (Singer, 2004a, 2004b, 2006a) news organizations.

The BBC study offers a good example of the detail that close examination of a single case can provide. Observation, interviews, and document analysis highlighted complex changes in professional status and skills, newsroom hierarchies, career opportunities, and traditional medium demarcations (Cottle & Ashton, 1999). Importantly, the newsroom fieldwork helped undermine the simplistic notion of technological determinism by incorporating issues of culture, environment, and situated experience, themes taken up subsequently by other researchers in this area, notably Boczkowski (2004a, 2004b).

Ethnographic approaches are useful for understanding "debates among organizational stakeholders about the most efficient, effective, equitable, and humane means of achieving their various goals" (Miller, Dingwall, & Murphy, 2004, p. 337). They are especially useful in exploring issues of process and explaining how outcomes are or are not achieved.

For journalists, motives behind convergence have been a central topic of debate. Many have seen it as part of a long-term corporate goal to cut costs related to the expensive news-gathering process. Newsroom executives have denied this motive, and indeed a lot of money has been spent on convergence—yet newsroom layoffs continue apace, and new hires are expected to possess multimedia skills. Few journalists are being directly compensated for additional work needed to create cross-platform content; instead, despite concerns of individuals and union leaders (Glaser, 2004) convergence-related tasks have been positioned as a new component of an existing job. Ethnographies have helped foreground the opposing views surrounding these issues. For instance, while one editor attributed an increase in published wire stories to greater reader interest in international news (Huang et al., 2004), a reporter described the bottom line of convergence as "fewer reporters, less real news gathering" (Friend & Singer, 2007, p. 216).

An ethnographic approach also has enabled diverse views to emerge about the core journalistic value of public service—and highlighted management's strategy of positioning convergence as a way to better serve an audience seeking news on demand from multiple sources. It thus has offered insights into stakeholder negotiations over a deeply contested issue: the stability of underlying principles amid significant structural and social change. "If [convergence] is about economic efficiency, then it isn't ever going to take hold in the newsrooms. If it's about quality journalism and doing things better with more tools, then it will," one manager said (Singer, 2006a, p. 40). Interviews with journalists indicated most had bought this argument that convergence facilitated expression and even expansion of their public service role; relatively few expressed concern about the potential for news monopolization (Singer, 2006a).

Finally, *ethnography and journalism bear a close relationship*, most obviously through their shared information-gathering methods of observation and interview but also in more complex ways. Denzin (1997) argues that ethnography involves storytelling that calls into question the nature and use of fact and fiction, issues with which journalists have long wrestled. Allen (1994) explores the idea of "media anthropology" as a way for journalists to communicate richer and more holistic narratives to their audiences, a capability facilitated by expanded media platforms.

The richness of interview data from ethnographic studies of newsroom convergence confirms that the method is one with which journalists are especially comfortable. Having spent careers fine-tuning their ear for a good quote, they know well how to deliver one. Could there be a pithier summation of the culture clashes between journalists in print and television newsrooms than this: "They've got the blow-dryers, we've got the investigative reporters" (Singer, 2004b, p. 844)? Or this description of the difficulty of negotiating culture-specific linguistics, offered by a television news producer who became a newspaper editor: "You can be the slot. It's a noun. It's a verb. It's an adjective. It's everything. And I have no idea what the hell it means. I know it's the cheese on the copy desk. Beyond that, I have no idea" (Silcock & Keith, 2006, p. 616). Or this insight into the life of a multimedia journalist and the potential toll on reporting: "You're filing for TV, you're filing for the radio, maybe 24-hour news want(s) to talk to you. When the bloody hell do you sit in the court and actually cover the case" (Cottle & Ashton, 1999, pp. 36-37).

As the examples suggest, ethnographic studies typically accommodate a narrative writing style that foregrounds individuals and their views in ways that are far more difficult with quantitative methods—much as interpretive or analytical journalism accommodates an authorial stance that goes beyond the detached objectivity of straight news reporting. Unbounded storytelling spaces afforded by the Internet also give journalists greatly expanded options to tell stories in meaningful ways. Indeed, as instant information becomes ubiquitous and as journalists are called on to more richly contextualize the news—to shift from gatekeepers to sense-makers—new outlets for ethnographic interpretation arise. It is hard to think of a topic for which the needs and interests of researcher and research subject so thoroughly overlap and inform one another.

Uses of Ethnographic Methods to Study
Newsroom Convergence

This section offers an overview of research into newsroom convergence over the past decade, much of it cited earlier but revisited here with a focus on methodological value. First are several key studies that have relied on qualitative fieldwork. They are followed by a quick look at surveys and content analyses, then by additional work that has combined traditional ethnographic methods with quantitative data collection.

Qualitative Ethnographies

A number of newsroom convergence studies have drawn exclusively on traditional ethnographic methods such as participant observation and in-depth interviews. Those summarized here are notable for their demonstration of the strengths of the method.

One of the earliest explorations of journalists' responses to the transition from a traditional, single-platform newsroom to one that demanded creation of content for multiple outlets was an ethnographic study of the BBC Newscentre in Bristol, published in 1999. It highlighted the ways in which news production technologies are culturally shaped by—and embedded within—corporate and professional contexts and practices. The researchers' assessment was that the transition caused additional pressures and frustrations for BBC journalists but overall had little effect on news output. Rather, journalists' existing approaches to news, derived from traditional newsroom culture and their socialization within it, were driving their actions in the new environment. "Despite the professional turmoil generated by the pressures and new working practices of multi-skilled, multi-media production, the news appears pretty much business as usual" (Cottle & Ashton, 1999, p. 41).

A few years later, Lawson-Borders (2003) focused on negotiations among organizational stakeholders in converging newsrooms. Drawing on literature from innovation management and diffusion of innovations theory, she used participant observation, in-depth interviews, and document analysis from news organizations in Chicago, Dallas, and Tampa to develop a seven-pronged framework of "best practices." She highlighted the value of allowing elements to emerge from the research rather than predefining the relevant concepts; her categories of best practice, from an emphasis on cooperation to the centrality of the audience member, stemmed from the data rather than being imposed in advance.

Boczkowski's (2004a, 2004b) work highlighted the value of intensive study of a small number of cases. In selecting news organizations in Texas, New Jersey, and New York, he sought cases that differed in important ways in their approaches to multimedia and interactivity, the dimensions of particular interest. He immersed himself in these newsrooms, spending four to five months on each case and racking up 700 hours of observation, 142 interviews with subjects representing various hierarchical levels and occupational roles, and analysis of a wide range of documents. The resulting publications (Boczkowski, 2004a, 2004b) show how occupational and organizational environments affect journalistic products and processes in unique ways: Terms such as "convergence" are linguistic shorthand for a great many activities that affect, and are affected by, specific individuals.

Silcock and Keith (2006), studying converged newsrooms in Tampa and Phoenix, were concerned with issues of culture in general and language in particular. They sought to understand not only what convergence means to journalists involved in it but also how participants negotiate the difficulties of communicating without a shared cultural vocabulary. Using a snowball sampling technique to select newsroom informants who could provide information-rich data, they found the journalists, in effect, compromised: They adopted a few words of each other's jargon without necessarily understanding the terms fully. Like Boczkowski, and as suggested by Dailey and his colleagues (2005), they found considerable variation within the umbrella concept of convergence, stemming from cultural differences at individual organizations (Silcock & Keith, 2006).

Quantitative Studies of Newsroom Convergence

Quantitative work complements the ethnographic studies of primary interest here. For example, Ketterer and his colleagues (2004) used a content analysis to examine the effects of convergence on the content of previously competitive print and broadcast media outlets in Oklahoma City. They found that despite corporate claims and branding efforts, there was little real cooperation between the two organizations. Nor was there evidence that the partners were meeting their stated goal of increasing in-depth news coverage.

Several U.S. surveys have addressed the cultural issues of newsroom convergence with which straight ethnographies are chiefly concerned. The widely cited "convergence continuum"—which posits a range of activities from simple cross-promotion to full convergence, the stage at which partners draw on the strengths of each medium to cooperate in planning, gathering, and disseminating the news—was based on surveys of newspaper editors and television news directors (Demo, Spillman, & Dailey, 2004/2005). A survey of newspaper executives by Bressers and Meeds (2007), focusing on the convergence of print and online operations, highlighted the importance of including online staffers at daily news planning sessions, as well as use of a central news desk (which few news operations actually had) to handle stories for multiple platforms (Bressers, 2006).

In another national survey, Huang and his colleagues (2006) found that nearly 40 percent of U.S. journalists believed convergence would lead to declining quality—but the same number thought it would not. Filak (2004) found that print journalists saw their professional culture as superior to that of broadcast journalists—and broadcast journalists similarly saw their own culture as superior. He emphasized the need for both groups to be involved in convergence planning to minimize likelihood of its rejection on cultural grounds.

Triangulation: Blending Approaches within an Ethnographic Framework

Triangulation is a process of using multiple perceptions to clarify meaning and identify different ways of seeing a phenomenon. A number of convergence studies have triangulated methods to enrich the understanding of this complex change.

An early study focused on journalists before they got to the newsroom: Three Brigham Young University educators used their school's efforts at curriculum revision to prepare students for multi-platform careers as a holistic case study, drawing on their own experiences, student interviews and questionnaires, and associated documents. The converged curriculum was met with mixed reactions; while some students liked it, other sentiments ranged from "confused" to "miserable" (Hammond, Petersen, & Thomsen, 2000, p. 24).

Huang and another set of colleagues (2004) combined a content analysis with an in-depth interview in a case study of the quality of journalism following newsroom convergence at the pioneering *Tampa Tribune*. They found that three years into its convergence experiment, the paper had not lost quality—but also that *Tribune* journalists were not engaged in significant amounts of cross-platform reporting. Most of the convergence efforts involved sharing tips and information, as well as cross-promoting news partners.

European scholars also are drawing on multiple methods to understand newsroom convergence. The centerpiece of Meier's study of convergence and innovation in three central European countries was a case study of the Austria Presse Agentur in Vienna that combined in-depth interviews of a dozen editorial staffers with an online survey. He found interactions between reconfigurations of the physical workspace and changes in both the speed and quality of journalists' work (Meier, 2007). In the first stage of their newsroom convergence project, a group of 25 Spanish scholars used interviews with editors to complement data from media and corporate websites. They plan to conduct ethnographic observations and consider the content contributions of both professional journalists and audience members (Domingo et al., 2007).

All three of my published convergence studies were based on triangulated data, drawn from newsroom observation, in-depth interviews, and a questionnaire distributed to interview participants. The following section considers how this multifaceted approach contributed to interpreting and understanding the issues of interest.

Theoretical Configurations

So far, this discussion of newsroom convergence has paid little attention to the ways in which ethnographic data can contribute to theory-based understanding of the phenomenon. While ethnographic work need not be concerned explicitly with theoretical development, instead focusing on description and explanation (Hammersley & Atkinson, 1995), it can be of value in exploring broader concepts. Incorporation of the sort of quantitative approach usually associated with empirical theory testing is helpful in this endeavor. Triangulation not only helps guard against seeing what's not there, a potential bias of any single-method approach, but also facilitates seeing what *is* there by enabling the researcher to go back and forth between distinct but complementary data sets.

In the winter of 2003, with support from the University of Iowa and the Kappa Tau Alpha journalism honor society, I spent a week in each of four newsrooms—in Dallas; Tampa; Sarasota, Florida; and Lawrence, Kansas. Each newsroom differed from the others in its approach to convergence and its organizational structure (Singer, 2004a, 2004b, 2006a). In addition to participant observations, I interviewed 120 journalists and newsroom executives; I also obtained 90 completed questionnaires from interviewees who were working journalists. The questionnaire covered the perceived impact of convergence on careers, work routines, public service, and the profession of journalism. Room for open-ended comments allowed subjects to highlight points important to them, frequently reiterating or supplementing themes from their interviews.

The triangulated data suggested three broad themes of conceptual interest. First, the portrait of convergence that the journalists painted through their words and questionnaire responses offered almost textbook illustrations of key elements in diffusion of innovations theory as outlined by Rogers (2003). These included characteristics of the innovation itself (the idea of convergence as well as procedural, structural, and cultural changes within the newsrooms), the communication channels that journalists were using to negotiate the change, and the social structures facilitating or hindering it (Singer, 2004b).

Journalists described a number of relative advantages of convergence over traditional approaches to newsgathering and dissemination, including perceived benefits for their news organizations, their own careers and their audiences. However, each set of data also indicated concern related to other innovation attributes such as compatibility (notably ongoing culture clashes and competitive issues) and complexity (expressed largely in terms of a desire for more training).

Other triangulated findings supported the importance of interpersonal communication channels to successful newsroom convergence, as diffusion theory would suggest; provided tentative evidence for a theoretically predictable adopter curve and for the presence of well-integrated newsroom opinion leaders; and indicated identifiable effects of physical, management, and social structures, again as suggested by diffusion theory (Singer, 2004b). In general, the application of diffusion of innovations theory suggested probable success for newsroom convergence efforts; indeed, since the studies were conducted, cross-platform content production has become routine.

The qualitative and quantitative data were not merely complementary; together, they enabled me to see theorized components more easily than I could by using either interviews or questionnaires alone. For example, a number of journalists discussed the importance of working one-on-one with colleagues in partnered newsrooms; "proximity breeds collegiality, not contempt," one news manager said (Singer, 2004b, p. 13). But it was the questionnaire data that drove home the importance of this perceived compatibility for successful convergence. Of 54 items, the statement generating the highest level of agreement was "I enjoy working with people who have professional strengths different from my own"—not a single respondent disagreed! Not far behind was "I have gained respect for the people in other parts of the news operation as a result of convergence." This surprising finding was one I might have overlooked in the interview data alone had the questionnaire not highlighted its importance.

Because diffusion of innovations theory has been applied to a great many changes, its components provide a useful pattern against which a new innovation can be compared. A different framework, drawn from the sociology of news literature, suggested that something else also was going on: As the social system in which they worked changed, print journalists were being resocialized. That is, their perceptions were shifting so that they no longer saw themselves exclusively as "newspaper people" but as members of a broader category that encompassed former "others" who worked in television or online media (Singer, 2004a).

However, there was considerable ambivalence about this shift, and one of the greatest contributions of the combined quantitative and qualitative data was to tease out the mixed feelings. For example, the questionnaire statement generating the second-highest level of agreement was "The quality of the story is more important than the technology used to tell it." But subjects were individually unsure and collectively divided on whether convergence produced more or less quality. They also had mixed feelings about competition and fast-paced information delivery. And while almost all said in their interviews that they wanted more training with new story-telling tools and techniques, they were generally neutral on the statement "The need to produce stories in multiple formats stresses me out." One converged reporter said he was at first scared he would say something stupid on television, but now "my heart rate barely picks up. My biggest fear is that my lips are chapped" (Singer, 2004a, p. 849).

Such mixed results, foregrounded by the use of multiple methods, underscored the fact that a change this significant is difficult for those in the middle of it, who must wrestle not only with the shift itself but also with their own reactions and perceptions. In other words, triangulation highlighted issues of culture, as experienced by participants, which are of integral concern to ethnographers.

A third published work focused on the normative concept of journalism as a form of public service and the ways in which journalists saw convergence affecting that norm. Although all journalists emphasize public service, demands of different formats result in different pressures for news workers; convergence challenges the "us" and "them" view taken by practitioners in a particular media domain (Singer, 2006a).

In the two publications discussed above, I started with a theory (diffusion) or conceptual framework (resocialization) and interrogated the data for relevant insights. For this third piece, I conducted a more structured discourse analysis of the interview data, extracting references to various aspects of public service and grouping the material into normative categories such as "accuracy" and "independence." Questionnaire responses supplemented this analysis by offering various articulations of categorical concepts, as well as more general deconstructions of public service. Examples of the latter include such statements as "My company is better able to serve our audience because of our decision to converge news operations" (most journalists agreed) and "My company converged newsrooms in order to do a better job providing information to various audiences" (journalists agreed again, though less strongly).

Although the published article focused on interview data, the questionnaires served as a crucial safeguard against the potential danger that findings will reflect the method of inquiry in misleading ways. The analysis suggested that within these newsrooms, convergence had failed to raise serious concerns about its fundamental compatibility with the core journalistic norm of public service. Although journalists worried about counterparts in partnered media doing things in different and potentially problematic ways, most supported convergence as an appropriate activity for a news organization committed to public service (Singer, 2006a).

Summary and Conclusion

This chapter has used newsroom convergence as a case study for the value of ethnographic research, broadly defined to incorporate not only traditional qualitative methods such as participant observation and in-depth interviews but also quantitative approaches such as questionnaires. It has highlighted core attributes of ethnographic work and shown how researchers have drawn on those attributes to study this multifaceted change in professional culture and practitioners' perceptions about it.

The changes are ongoing. Not only does nearly every media outlet now have an online presence, but industry data suggest that more journalists are creating cross-platform content than was the case even a few years ago, when most of the academic work cited here was conducted. News organizations as venerable (and tradition bound) as *The New York Times* have combined their print and online news operations. Most major newspapers offer blogs by journalists on their websites. Forty percent of recent hires report that their jobs include writing and editing for the website, with the numbers who are designing and building web pages also rising (Becker, Vlad, & McLean, 2007). "The pace of change has accelerated," the Project for Excellence in Journalism (2007) confirms. Moreover, "the transformation facing journalism is epochal, as momentous as the invention of television or the telegraph, perhaps on the order of the printing press itself."

Ethnography will continue to be an optimal method for exploring the nature and effects of this enormous cultural transition for journalists and journalism. It is ideally suited to understanding not just causes or effects, not just products or practices, but also the processes that underlie them, the perceptions that drive and are driven by them, and the people who have always been at the heart of the journalistic enterprise, whatever its iteration. Importantly, those people are not just journalists but also members of the public, "the people formerly known as the audience" (Rosen, 2006). The idea of convergence needs conceptual expansion to include the mingling of the roles and the products of journalists and nonjournalists within an open, networked digital environment (Domingo et al., 2007; Quandt & Singer, in press). As that network becomes an increasingly dominant communications medium, the need to understand the fluid interactions and interconnections it generates will steadily increase. The questions asked will be cultural ones in every sense of the word, and ethnographers will play a central role in addressing them—and asking interesting new ones.

CHAPTER ELEVEN

The Active Audience: Transforming Journalism from Gatekeeping to Gatewatching

Axel Bruns

Axel Bruns is a vocal advocate of the democratizing potential of citizen journalism and the power of bloggers; his dismissal of traditional media providers in the face of a new model of journalism is provocative, and if entirely accurate, might render meaningless the efforts of old media to embrace the new. In a departure from the ethnographic narratives of other chapters, we include Bruns' analysis of political blogging to expand our coverage of the changing nature of journalism. The theoretical proposals of the chapter are complemented with methodological suggestions that hint at the future challenges for online journalism research.

EDITORS' NOTE

In what to many Australians was an interminably slow lead-up to the federal election held in late 2007, an unusual spectacle played out across the pages of one of the leading national newspapers, *The Australian,* and a multitude of news, current affairs, and commentary weblogs and citizen journalism websites. A publication of News Ltd., the domestic arm and foundation stone of Rupert Murdoch's NewsCorp empire, *The Australian* has long positioned itself as a loyal supporter of the incumbent government of Prime Minister John Howard, and is widely regarded as generally favoring the conservative side of politics. It continued to do so even in the face of opinion polls (some by News Ltd.'s own polling agency, Newspoll) which throughout 2007 consistently showed both a commanding lead for the opposition Labour Party over the conservative Coalition government on a two party-preferred basis, and a strong preference for opposition leader Kevin Rudd as Prime Minister.[1]

The Australian's commentators sought hope amidst the conservative despair; however, small upward movements in poll results in favor of the government were described more often than not as another sign that "the honeymoon is over" for the opposition leader, while movements in the opposite direction were explained away with references to the polls' margins of error. Australian political bloggers and citizen journalists, meanwhile, found great pleasure in analyzing and critiquing such commentary (and, to a somewhat lesser extent, that of other newspaper and broadcast journalists): dubbed the *Government Gazette*,[2] *The Australian*'s editorial pages and their online counterparts were examined and found wanting on an almost daily basis. As many of the paper's editorial pieces were also published on News Ltd.'s *News.com.au* website (combining material from *The Australian* and other Murdoch papers around the country) and there featured direct commenting and discussion functions for users, a significant amount of such criticism also found its way onto the *News* website itself, displayed immediately alongside *The Australian*'s editors' and commentators' opinions.

Ultimately, the persistence and vigor of such grassroots criticism appeared to have a surprisingly strong impact: on July 12, 2007, the paper published an extraordinary article[3] openly attacking bloggers and other "sheltered academics and failed journalists who would not get a job on a real newspaper," ostensibly for daring to voice their disagreement with *The Australian*'s own journalists' and pundits' interpretation of the political mood of the electorate. The article denounced grassroots online commentators as "out of touch with ordinary views," and culminated in the remarkable assertion that "unlike [political commentary site] Crikey, we understand Newspoll because we own it."

The Decline of Gatekeeping

This meltdown at the *Government Gazette* points to a larger challenge for the mainstream journalism industry. "A.J. Liebling once said, 'Freedom of the press is guaranteed only to those who own one.' Now, millions do" (Bowman & Willis, 2003, p. 47; original quote: Liebling, 1960); similarly, while those millions who are now active as part-time news bloggers, citizen journalists, and political commentators do not (yet?) operate their own opinion polling services, there is little to stop them from offering their own interpretations of available polling data, informed by their own experience as average citizens rather than by the accepted wisdom of professional journalists or the political spin of media minders. Indeed, at least in the Australian example, some of the most insightful analyses of polling data are available from blogs such as *Poll Bludger* or *Mumble*, operated by student and professional psephologists—scientists specializing in the statistical analysis of voting intentions and election results. (One of them, *Mumble*'s Peter Brent, was the only blogger mentioned by name in the editorial, having been informed by phone the previous day that the paper would "go"— Australian slang for 'attack'—him.)[4] Especially in such cases, the expertise of nonjournalists in reading the mood of the populace clearly exceeds that of journalists and pundits, regardless of who owns the polling services themselves.

What *The Australian*'s tirade against citizen journalists points toward, then, is a deep-seated and justified concern amongst industrial journalists (well beyond Australia) that their new grassroots counterparts have begun to undermine mainstream journalism's traditional position of influence and importance. Once able to lead—indeed, to form—public opinion, papers such as *The Australian* now appear hardly able to follow or comprehend it; those whom Jay Rosen (2006) has referred to as "the people formerly known as the audience" have begun to look elsewhere for news and informed opinion, or have begun to create and publish their own reports, commentaries, debates, and deliberations on news and current affairs, especially in the online environment. To some extent, this shift is one of journalism's own making, as the industry's failure to update its products for a new, Internet- and convergence-driven environment has alienated younger audiences, and as many journalists' inability to remain politically and commercially independent has been highlighted by the utter failure of mainstream journalism in the United States, Australia, and elsewhere to debunk the unsubstantiated Weapons of Mass Destruction claims used as a pretext to start the war in Iraq, by cases of preferential treatment for major advertisers, and by the growing conflation of news and entertainment content especially in television broadcasts (Lowrey & Anderson, 2005). Newspaper readership and credibility has fallen to record lows (Project for Excellence in Journalism, 2007); indeed, at least in the United States, many appear to prefer to receive a good part of their news in the form of the pithy news satire provided by Jon Stewart and Stephen Colbert (see Fox et al., 2007).

On the best evidence currently available, there is little indication that this trend in print and broadcast news is likely to be reversed any time soon; like other informational industries from software to music, the news industry in print and broadcast is operating under a business model which no longer suits the emerging cultural and economic framework (Jenkins, 2006). Both print and broadcast proceed from an industrial logic which is founded on the twin assumptions that their means of production are expensive and concentrated in the hands of a small number of major operators, and that access to their channels of distribution is tightly policed and therefore scarce; neither assumption, however, still holds true in a postindustrial, Internet age. Today, textual as well as, increasingly, audio and video content can be produced and widely distributed at a negligible cost by a very broad range of participants; further, both production and distribution can be organized through the harnessing of collective and distributed activities just as effectively as it has been, traditionally, through centralized, corporate efforts. Not only, then, do traditional print and broadcast journalism operators face new competition from the online journalism industry (which, for the most part, still has to find its own models for sustainable revenue generation); industrial journalism as a whole is now also in competition with an almost entirely new group of collaborative, citizen journalism projects.

It becomes all the more important, then, to develop research methodologies which provide clear quantitative as well as qualitative insight into the makeup, agendas, and models of operation and collaboration in citizen journalism and the wider news and political blogosphere—and ethnographic approaches play an important role in this context. Work done in this field to date remains limited, however—both because of a lack of established research tools and methodologies, and because of a limited understanding of how we may define citizen journalism itself. Before we explore possible avenues for research, then, let us theorize the practice of citizen journalism.

In part, citizen journalism remains a somewhat nebulous concept also because some industrial online journalism operators have been quick to claim their sites as "citizen journalism" for little more than the fact that they offer on-site forms for audience responses. Beyond such window dressing, however, citizen journalism is described more appropriately as a form of journalism where citizens themselves, rather than (or at the very least in addition to) paid journalists claiming to represent the public interest, are directly engaged in covering, debating, and deliberating on the news. Such sites break with the traditional logic of journalistic operations (justified itself by the concentration and scarcity of the means of production and distribution in industrial journalism), by which journalists and editors play a crucial part in the preparation and selection of those news reports which are ultimately presented to audiences as "all the news that's fit to print" (or to broadcast).

That logic introduced three bottlenecks into the news production process: one at the input stage, where editors and journalists make a preliminary selection of what current and upcoming news events may be worth covering for the next edition of the newspaper or broadcast bulletin; one at the output stage, where editors pick the final selection of articles or reports to be included in the publication; and one at the response stage, where editors again select a small sample of reader or viewer comments to be presented on the Letters to the Editor page or in call-in radio or TV shows (fig. 1).

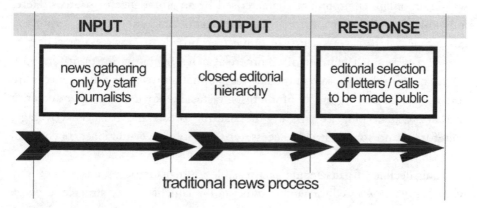

Figure 1: Stages of the traditional news process

Each of these bottlenecks is justified in journalistic tradition by noting the scarcity of the print or broadcast medium—where the newshole (the total amount of column inches or broadcast minutes available for covering the news) is strictly finite, it clearly is of the utmost importance that the content which does make it into the final product is tightly policed to be as relevant for the intended audience as it can possibly be. This process of selection, known as gatekeeping (see McQuail, 2005, p.308; Shoemaker, 1991), remains one of the fundamental activities of any journalist or editor, taking place as it does at each of the three stages (input, output, and response) as well as at levels ranging from that of the individual article (where ideally, gatekeeping ensures coverage of the core facts and excludes less relevant information) to that of the publication as a whole. As a process of selecting what events to observe, what stories to cover, and what responses to publish, the quality of gatekeeping necessarily depends on the expertise of staff (and on their independence from commercial or political interference), and is tied directly to the availability of time and resources sufficient to enable considered selection decisions to be made; in news organizations where gatekeeping decisions are largely devolved to junior staff or outsourced to transnational wire services, the diversity and quality of news as well as the relevance of news to local audiences is likely to suffer.

At the same time, while in a print and broadcast environment with a small number of news producers it was incumbent on journalists to ensure they provided as broad and comprehensive a coverage of news events as possible (to offer if not literally "*all* the news that's fit to print" then certainly the most important news across all beats), this obligation no longer applies in the same way in an online environment characterised by an abundance of sources. Here, individual sources are freed from the public responsibility that comes with being granted access to a very scarce resource (the broadcast airwaves) or acting as one of a very small number of commercial news operators (in print); instead, online news outlets are safe in the knowledge that a myriad of other sources are sure to cater to a wide variety of interests, tastes, ideologies, and other audience markers. Gatekeeping as a means of ensuring broad and balanced coverage, therefore, is no longer strictly necessary; the gates have multiplied beyond all control.

This decline of gatekeeping as a feasible journalistic practice is also exactly what *The Australian*'s editors and commentators lament as they struggle to come to terms with their new citizen journalist competition. Operating the gates of journalistic publications conferred a significant degree of power on editors and journalists, allowing them to direct the public gaze toward or away from specific news topics and placing them in a position to act as opinion leaders; the more readers and viewers have transformed into browsers, users, and even what can be described as *produsers* of news reports and commentary (Bruns, 2007a,b), the more this power is being wrested away from the mainstream news media. In this light, *The Australian*'s editorial must be read as a groan of frustration at this newfound impotence, and its assertion that only "we understand Newspoll because we own it" becomes a pathetic last claim to an authority based on exclusivity that is now fatally undermined.

Gatewatching

What has emerged as an alternative to gatekeeping is a form of reporting and commenting on the news which does not operate from a position of authority inherent in brand and imprint, in ownership and control of the newsflow, but works by harnessing the collective intelligence and knowledge of dedicated communities to filter the newsflow and to highlight and debate salient topics of importance to the community. The community of bloggers, citizen journalists, commentators, psephologists, activists, and others which so taunts and torments *The Australian*, for example, does not aim to supplant it or any other newspaper or online news site by developing its own comprehensive news service; what it does do is to offer a corrective, an alternative interpretation of the day's events, and to round out industrial news and other sources by adding the backstory and providing further related (and often contradicting) information enabling readers to better assess for themselves and by themselves the quality and veracity of mainstream news stories, press releases, reports produced by the government and nongovernmental organizations (NGOs), opinion pieces, and other material as it becomes available.

The net result of such practices is to help fellow users make sense of this avalanche of reports, commentary, and opinion in both commercial and noncommercial online news and information sources. This is decidedly not a matter of gatekeeping in any traditional sense of the word: Audiences now have direct access to a multitude of sources, and no longer rely on journalists to report the statements of politicians and other public actors, the news releases of governments and corporations, or the opinions of pundits and commentators. Instead, it is a matter of *gatewatching*, of observing the many gates through which a steady stream of information passes from these sources, and of highlighting from this stream that information which is of most relevance to one's own personal interests or to the interests of one's wider community (Bruns, 2005). Of course it should be noted that some such gatewatching was practiced already in traditional journalistic industries, too; here, journalists intently watched the gates of government and corporate organizations as well as of the news agencies to which they subscribed (not to mention those of competing news outlets), to identify any potentially newsworthy material to be fed into the subsequent *gatekeeping* process. In the online gatewatching environment, however, agency has shifted from the journalistic profession to anyone interested in getting involved in the process—from individual news bloggers highlighting, in more or less "random acts of journalism" (Lasica, 2003), those news which speak to their personal concerns, to the many citizen journalism communities collaborating to specifically gatewatch news sources within their field of shared interest.

This collaborative gatewatching effort differs from traditional journalistic practice in a number of significant ways. On the one hand, at the input stage, it clearly foregrounds the information discovery effort inherent in gatewatching over the information summary effort of traditional journalistic writing; rather than synthesising multiple sources into one coherent news report which (after a gatekeeping stage aimed at shaving off any apparently extraneous material) is then published as a product in itself, gatewatching merely compiles one or a number of related reports on a newsworthy event, thereby *publicizing* the event and the stories which cover it rather than *publishing* a news report. On the other hand, the gatewatching process is not at all complete at this stage; it typically remains open for users to add further gatewatched information during the response stage as well, enabling an ongoing coverage of the event even beyond the initial report. Through such responses, the initial reports are fleshed out, examined, critiqued, debunked, put into context, and linked with other news, events and background information; this process externalizes and turns into a widely distributed collaborative effort similar processes which previously took place either entirely within the minds of active news consumers, or within small, relatively isolated groups of consumers discussing the news of the day.

Gatewatching as we have described it here can today be seen in process across a wide range of online news sites. It is found in the technology news site *Slashdot* as much as in the operations of many individual and group news bloggers; it takes place in collaborative pro-am news experiments like *Oh-myNews,* which combines both volunteer citizen journalists and professional editors, and even in the automated gatewatching of *Google News.* How these sites manage their operations differs from case to case, and it is important to examine the implications of these differences (as I have done in Bruns, 2005)— some sites operate as a free-for-all where all gatewatched content is published to the wider Web immediately; some combine their gatewatching of external sources with a kind of internal gatewatching process which enables their community to collaboratively highlight from all incoming submissions those reports which are seen, communally, to be of the highest importance; some even retain a small degree of internal gatekeeping as a means of sorting through the material discovered through gatewatching. None of these models are inherently better or worse, of course; the appropriate model must be chosen as suitable to the communities and contexts it is meant to serve. What is import-ant, at any rate, is that the gatewatching model departs crucially from the gatekeeping model employed traditionally by the journalism industry (fig. 2).

As a result, news turns from a relatively static product to be consumed by audiences into a dynamic, evolving, expanding resource that is actively co-developed by the users of such citizen journalism sites, participating a process of *produsage*. Such news produsage follows the same logic which has seen the active co-development of software under open source models or the collaborative co-creation of encyclopedic resources on the *Wikipedia*, and here as well as there significantly affects industry incumbents. If, as has been pointed out in a number of contexts, there is an overall shift from passive consumption to active participation in our societies (see Jenkins, 2006; Benkler, 2006; von Hippel, 2005), then this poses a key challenge to a journalism industry built traditionally on a conception of its audience as passive consumers, and—as *The Australian*'s editorial documents—most of the industry has yet to develop strategies for addressing this challenge effectively.

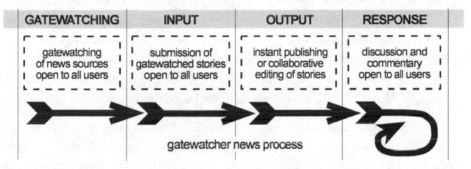

Figure 2: The gatewatcher news process

Some industry responses to the emergence of citizen journalism have been predictable, much like the software and encyclopedia industries. For example, the journalism industry worldwide has repeatedly attempted to denounce its citizen counterparts as unskilled amateurs providing low-grade and untrustworthy content; such criticism is undermined, however, by the recent history of significant failures in industrial journalism itself. More progressive elements in mainstream journalism have attempted with varying degrees of success to incorporate elements of citizen journalism into their own operations; this has ranged from the mere provision of more or less tightly policed online discussion fora for readers on *News.com.au* to the *L.A. Times*' failed experiment with enabling its readers to revise its daily editorials, wiki-style (Glaister, 2005). Such experiments are frequently undermined, however, by the already poor relationship between news corporations and their readers (who regard most attempts to embrace the citizen-reader as little more than cynical window dressing), and a common failure of news organizations to change journalistic attitudes at the same time as they are changing news technologies (Chung, 2007; Hermida &

Thurman, 2007). The attachment of blog-style elements (such as commenting and discussion functions) to online op-ed columns, for example, is of little consequence if op-ed pundits maintain a generally condescending attitude towards their audience or fail to respond to them altogether; indeed, it only serves to highlight the growing detachment of the punditariat from overall public opinion which is being observed in a number of Western nations.

Instead, journalism would do well to re-imagine its audience and reinvent its professional practices, much in the same way that parts of the software industry already have in the wake of open source: here as well as there, there still remains an important place for the industry professional, but their role now is one of guide rather than leader, of (complementary) service provider rather than (sole) content producer—professional journalists can participate most effectively when they contribute original research and promote public debate, rather than acting as gatekeepers to summarize, contain, and conclude public debate.

Researching Citizen Journalism

This, then, lends even more urgency to the need for further research into the wider ecology of journalism—an ecology which now irrevocably contains both industrial and citizen journalism, existing in a variety of adversarial, parasitic, or symbiotic relationships. The research necessary in this field spans a wide range of approaches from the local and specific to the global and generic, and from the qualitative to the quantitative; it must examine the conduct of individual participants and publications as much as it needs to provide insights into the interconnections and patterns of information flow between individual sites and between their communities of users. A variety of research methodologies are now being tested to achieve such aims, even if few have as yet risen to signifi-cant prominence in the field—this is hardly surprising, perhaps, given the still highly fluid nature of citizen journalism communities and practices, as well as the continuously changing and evolving technological frameworks upon which they are built.

In the first place, direct ethnographic studies of participants in specific citizen journalism sites and projects will yield important information about their practices and motivations. In this context, it will be important to avoid overgeneralisation, and to distinguish clearly between different modes and levels of participation—bloggers only tangentially concerned with discussing news or current affairs engage in citizen journalism in a vastly different way from those participants who are regular contributors to a *Slashdot* or *Kuro5hin*, for example, and cultural and political contexts for such practice must also be recognised: participants in hyperlocal U.S. citizen media sites (see Schaffer, 2007) are likely to bring vastly different attitudes to their content production than do the citizen journalists registered at the Korean *OhmyNews*, or the pro-democracy bloggers operating under the radar in a number of Middle Eastern nations (see, e.g., Goussous, 2007) or in China (see chapter 8).

Broader ethnographic and demographic studies—including for example the work done at the Pew Internet & American Life Project—may also help to address another crucial question concerning the real impact of blogging and citizen journalism on wider public perceptions. While many mainstream journalists are clearly irritated by the rise of alternative, citizen-led media forms, do such reactions accurately reflect the role of citizen journalism in changing the overall mediasphere, or do they stem from a more general frustration with the decline of professional journalism? Put simply, do news blogs matter as much as journalists think (or fear) they do? A continuing growth in blogging and blog readership (and in access to and participation in citizen journalism more specifically) does appear to support that belief, but such figures alone may only indicate interest, not impact. Indeed, such research points to a broader question about how the public sphere may be reconfigured in a networked age in which the influence of conventional mass media has declined markedly.

Many answers to such questions may also be found through what has been labeled virtual ethnography (Hine, 2000), and through other, related approaches to content and network analysis. Such research has been boosted significantly in recent years with the development of increasingly sophisticated tools for the automated gathering of large data sets containing the footprints of participant interaction in blogs, citizen journalism websites, and the wider Web, but clear methodologies for the analysis of such data are in many cases yet to be established (Hine, 2005). Such research builds on the fact that in the online context, participants are represented solely by their utterances, most of which in the context of citizen journalism are publicly visible and highly interconnected, providing a rich resource of textual data linked to individual contributors and discursive communities (at the same time, however, a focus on the online dimension necessarily misses any offline characteristics of individuals and groups, and would therefore also benefit substantially from being complemented by further ethnographic study using more conventional methods).

Sources of information for such research include the utterances themselves (including relevant metadata such as URLs, date and time of publication, peer quality ratings, and further contextual information), the hypertext interlinkages between individual pieces of content, as well as more generic information such as the Google PageRank or Alexa rankings, or the Technorati authority scores, of the pages or sites on which the content was published, or the level of annotation through del.icio.us and similar social bookmarking services (Golder & Huberman, 2005). Using a variety of tools from the content analysis software Leximancer to network analysis tools such as IssueCrawler, it is now possible to extract a variety of quantitative information from such data; this provides synchronic as well as diachronic, momentary as well as longitudinal insight into the patterns of interaction, the flows of information, and the processes of opinion formation in online environments. Tracking the take-up of memes (terms, topics, links) over time as well as tracing the networks of interlinkage between sites can provide an indication of current key themes in the news and a clear indication of influential "opinion leader" sites whose content is of great salience to specific interest communities (Bruns, 2007c); more longitudinal approaches can also chart more general trends in blogs and citizen journalism— for example, they offer an insight into the levels of consistency or churn amongst the so-called "A-list" of influential bloggers in a given field of interest (Kirchhoff, Bruns, & Nicolai, 2007).

Additionally, such work can identify clusters of closely interconnected interest communities in citizen journalism, the blogosphere, and the wider Web (most likely existing around the main opinion leaders), patterns of interconnection between related clusters, as well as peaks and troughs of authority and influence in the network; again, of course, this research may be further enhanced through more manual forms of content analysis and ethnographic study which would, for example, be able to divide overall clusters or networks according to specific demographic or ideological factors (see Hargittai, 2005, for an analysis of patterns of interlinkage between political bloggers from opposing political camps in the United States). Further research would also return to a more conventional ethnographic mode, for example, by engaging directly with key actors in the network to study their practices and motivations, and to identify the factors that contribute to their ability to gain and maintain a position of recognition and influence in the citizen journalism arena. This work would mirror similar ethnographic studies of newsroom practices and journalist attitudes as they have been conducted in the journalism industry during much of the second half of the twentieth century.

Once fully formed, research approaches as we have outlined them here will be valuable in their own right, as methodologies for the study of broad-scale online (and especially web-based) communication, interaction, and collaborative publishing well beyond the citizen journalism or overall journalistic field. More specifically, they will contribute significantly to the development of a theory of the networked public sphere—or perhaps more correctly, a theory to describe the structures of the networked environment that supersedes the conventional model of a mass-mediated public sphere which is now increasingly out of step with observable reality by tracing information flows and opinion formation on the network and thereby providing a counterpart to studies which investigate other aspects of the overall media ecology. Only in combination will such studies be able to accurately describe the full operations of mediated communication at local, national, and global levels.

In the first place, however, the major contribution of this research is to offer a clearer and more detailed picture of how news and political blogging and citizen journalism operate in practice, of what contribution individual participants make to the distributed collaborative project, and of what motivations they have for their actions (see chapter 12). Such information is of use to professional journalists and the journalism industry, of course, as it cuts through misperception, misunderstanding, and misinformation (as well as, in some cases, outright ignorance), and instead enables journalists and institutions who so choose to develop a less belligerent, proactive, and perhaps more cooperative model for dealing with citizen journalists—ultimately undoubtedly a more productive approach than simply to "go" them, or to patronize them by offering faux pundit blogs or operating citizen journalism sandbox environments. Additionally, such research also provides an opportunity for journalism educators to more directly address citizen journalism as an area in need of well-developed journalistic skills.

There now exists a continuum of journalistic models and practices which stretches from industrial gatekeeping to citizen gatewatching, with a strong pull at present perhaps towards the central, hybrid space; industrial journalism which continues to ignore or denigrate citizen efforts is increasingly left behind public sentiment, but citizen journalism which shows scant regard for journalistic ethics or professional conduct similarly undermines its own position as a credible alternative. Professional journalists no longer simply inform citizens of the news, but work with citizens in developing a shared understanding of the news, just as members of the public move from a position as informed (but perhaps passive) citizens to one as active, monitorial citizens (Jenkins, 2006, p. 208). Rather than indiscriminately dismissing any commentators expressing diverging views as "woolly-headed critics" indulging in "smug, self assured, delusional swagger," as *The Australian* editorial literally stated, this means that professional journalists must treat their audiences, their users, their newfound citizen-journalist peers, with the respect they deserve—and more fundamentally, it requires the industry to critically self-reflect on the factors which have led many of the people formerly known as the audience to create their own alternatives to mainstream news.

NOTES

1. See, e.g., the survey results of Newspoll (http://www.newspoll.com.au/index.pl? action=adv_search) and Roy Morgan Research (http://www.roymorgan.com/news/ polls/federal-polls.cfm).
2. See, e.g., M. Bahnisch's blog Lrvatus Prodeo: "Government Gazette Fights Back" (July 11, 2007): http://larvatusprodeo.net/2007/07/11/government-gazette-fights-back/
3. The article was titled "History a Better Guide than Bias." Available online at: http://www.theaustralian.news.com.au/story/0,20867,22058640-7583,00.html
4. As stated by Brent in his weblog (mumble.com.au).

CHAPTER TWELVE

The Routines of Blogging

Wilson Lowrey & John Latta

While initially dismissive and critical, journalists of the old media world are now seeking collaborations with established bloggers and other non-professional journalists. But while much has been said about their influence, the work practices of bloggers have received little empirical study (nor do they lend themselves easily to the classic techniques of media production ethnography). Blogging can be examined from the perspective of professional journalism to explore to what extent these information providers diverge from their professional counterparts in routines and values. Lowrey and Latta's in-depth interviews with political bloggers suggest that the more relevant bloggers become in terms of audience and influence, the more their production routines resemble those of professional journalists.

EDITORS' NOTE

It has become a part of the blogging folklore that news blogs have the power to fundamentally change the way individuals receive and relate to news, making the public more aware of divergent points of view and sources and less reliant on traditional news media (e.g., Gillmor, 2004; Rosen, 2006). Blogging advocates argue that bloggers, by serving as watchdogs, can make traditional news media more responsive and responsible to the people (Gillmor, 2004; Bruns, 2005; Kim & Blood, 2005; Rosen, 2006). But do bloggers really have a better grasp on the needs and preferences of "the people" than do other types of media producers? A major challenge faced by all producers of media content is understanding audiences, when in most cases, cultural producers have limited, if any, direct contact with them (Ettema, Whitney, & Wackman, 1987; McQuail, 1997, 2005). All acts of cultural production, from traditional journalism to film making and fine arts, rely on conventions to accomplish work while approximating a shared understanding with audiences and while working within social, cultural and economic constraints (Altheide & Snow, 1979; Becker, 1982;

McQuail, 2005). Such conventions typically become routines (Shoemaker & Reese, 1996; Pottker, 2003; Atton & Wickenden, 2005), and it seems likely that the production of blogging also necessitates conventions and routines.

Routines are the by-product of an uncertain, complex environment coupled with organizational constraints and needs. They help media producers reduce uncertainty and accomplish work, but they can become taken for granted and deeply naturalized, invisible to those who rely on them (Ryfe, 2006). As such, routines and conventions can become ends rather than means, and they can have a strong impact on the selection of events as well as on media form and content (Shoemaker & Reese, 1991; Lowrey, 2006; Ryfe, 2006). As cultural producers who have limited direct contact with audience (lurkers far outnumber those who actually respond), it seems likely news bloggers would have developed their own conventions and routines, and that these would shape blog content. And despite a general claim to be unorthodox challengers to mainstream media, there is evidence that news bloggers rely on some of the routinized production methods, practices, and values of mainstream journalism (Singer, 2005; Wall, 2006).

Blogs and Journalism

Though the great majority of blogs take the form of personal diaries rather than public information (Herring et al., 2007; Pappacharissi, 2006), around 11 percent of blogs contain political content (Lenhart & Fox, 2006), and politically oriented blogs are second to personal and family weblogs in readership (Haas, 2005; Lenhart & Fox, 2006). Blogging enthusiasts think of these blogs dealing with current events as distinct from mainstream news media, calling them "a new genre of journalism" (Wall, 2005), amateur journalism (Lasica, 2003) and "folk journalism" (Haas, 2005; Mortensen & Walker, 2002). Scholars have distinguished bloggers from traditional journalists by their independence from organizations (Lowrey, 2006), by the fact that they filter information individually and subjectively, and not institutionally (e.g., Gillmor, 2004; Haas, 2005; Kim & Blood, 2005; Matheson, in press; Singer, 2005), and by journalism's formalization of ethical principles (Singer, 2006b). Some scholars and blogging advocates say bloggers are already changing journalism by using "a more conversational, dialogic, and decentralized type of news" (Delwiche, 2004; Sullivan, 2004) that offers the public subjective, multiperspectival and intertextual content, in contrast to the more rigid and closed content of institutional news (Bruns, 2004;

Gallo 2004; Haas, 2005; Matheson, 2007). And in fact, researchers have found some evidence that journalists are adjusting practices to repair vulnerabilities brought to light by the blogging challenge (Lowrey & Mackay, 2006), and that bloggers are having limited influence on the media's agenda (Delwiche, 2004; Dylko & Kosicki, 2006).

However, bloggers tend to reproduce rather than challenge the work of mainstream media, and they may adopt similar practices (Haas, 2005). News-oriented blogs use the same topics and often borrow the same stories as a narrow range of elite traditional media (Delwiche, 2004; Haas, 2005; Murley & Roberts, 2005; Scott, 2006; Trammell & Keshelashvili, 2005). Even when seeking distinction, bloggers rely on traditional journalism as a position against which to define themselves (Matheson, 2004).

Bloggers most often add links in their stories to well-known blogs and mainstream news sites rather than to fresh, radical, alternative sources, and bloggers tend to rely on and support blogs that are familiar and similar (Haas, 2005; Singer, 2005). A study on blogging in the Iraq War found that existing frames about the war-dominated discourse (Wall, 2006), and other research has shown bloggers tend to monitor each other or the traditional media rather than directly report events in the world (Singer, 2006b; Scott, 2007). Research into networks and organizational ecology offers a possible rationale, showing that early entrants into a network or field acquire the most connections, and late entrants struggle for attention (Barabasi & Albert, 1999) and are more likely to fail (Baum, 2001).

Just as news producers such as *The New York Times, The Washington Post,* and *CNN* signify quality in the field of mainstream journalism, a few bloggers are considered elite, or "A-list," and serve as intermedia agenda setters for other bloggers (Haas, 2005; Singer, 2005). Bloggers may engage in "impression management"—sociologist Erving Goffman's term for the process of controlling impressions others have of us (Goffman, 1959). Bloggers try to position themselves as legitimate or superior within the blogging community (Trammell & Keshelashvili, 2005). The quality and quantity of the referrers of a blog determine success (Glaser, 2003): "He or she who gets the most links wins in the world of weblogging" (Powers, 2003).

However, the pursuit of status, audience and revenue may correlate with the homogenizing and mainstreaming of content, and the shedding of the ideological niche (Lowrey, 2006; Scott, 2007). Scott (2007) warns that this desire for financial reward fits the communication needs of the power structure: "One can easily see potential slippage of bloggers from scrappy, upstart challengers to apparatchik pawns of sophisticated strategic communication campaigns" (p. 52).

Bloggers may desire more readers, but like journalists, they do not meet or converse with most of their readers directly. In traditional news work, journalists tend to construct audiences to suit production needs—i.e., they develop audience routines—rather than letting audience interests and needs determine production (Ettema & Whitney, 1994; McQuail, 1997, 2005). Bloggers may do the same, particularly as they grow in popularity and audiences become more complex, as network theory suggests will happen to early entrants to the field.

If bloggers do adopt routines, how might they compare with traditional news routines? The routines of traditional journalism have been defined as repeated procedures, practices and forms that make it easier for journalists to accomplish tasks in an uncertain world while working within production constraints, and which tend to become naturalized (Lowrey, 2006). There are numerous examples of news routines. Journalists "typify" events into discrete news categories such as hard vs. soft, thereby gaining prediction and control over news flow and production in a complex world (e.g., Tuchman, 1978). The routine of objectivity requires balanced sources, personal detachment in reporting, and reliance on official sources, each of which helps news organizations gain wider audiences and avoid lawsuits, and in the process helps support those who benefit from the status quo (Soloski, 1997; Tuchman, 1978).

Alternative media that claim opposition to traditional media have also found they benefit from routines that reduce uncertainty. Even radically egalitarian media efforts such as IndyMedia have sought efficiency by pursuing shortcuts in decision making, some of which serve a latent hierarchy (Pickard, 2006). Reporters at alternative, politically slanted media select sources in part because they know the advocacy groups and their leanings (Atton & Wickenden, 2005; Patterson & Donsbach, 1996), and reporters risk management disapproval if they bypass the standard source list (Eliasoph, 1997).

It seems likely that bloggers rely on routines as well. By providing some understanding of the nature and context of routines in blogging, this study contributes another perspective for the general study of media routines.

A number of research questions are asked:

RQ1: What routines are most evident in bloggers' daily production work?
RQ2: How do these routines compare with the routines of traditional journalism, as discussed in the literature on news construction?
RQ3: What factors necessitate formation of routines in blogging?

Methodology

Six in-depth, semistructured personal interviews with current U.S. bloggers were conducted in spring 2006. Only blogs focusing primarily on social and political issues and events were selected, but variety was sought in range of intended audience: One blog focused on issues and events in the researchers' city (population approximately 100,000), one blog focused on particular neighborhoods in a nearby large city, one blog focused on politics in the researchers' state, one blog focused on national and state issues and another blog focused on national issues, with an African-American perspective. Four of the bloggers were male, and two were female. One blogger was African American (a male), and the others were white. The male respondents were in their 20s, one woman was in her 50s, and the other was in her 50s. Only one of the bloggers—a male—worked full time. Three were students, and two had part-time jobs.

Each interview lasted approximately one hour. Questions were open ended, and follow-up probe questions were used as well. Respondents were asked about constraints, challenges, and obstacles, as routines develop in response to these. Questions about audience and source information were asked in order to probe for conventions related to reducing uncertainty. Questions about fellow bloggers ("occupational peers") and other blogs were asked because available models for mimicry can shape routines. Interviews were grounded in the respondents' daily work experiences by asking them to recount daily production and by viewing the blogs during interviews. A set of initial questions was developed prior to interviews, and these were asked of each respondent, though the interviewer allowed respondents to stray down tangential paths when they promised insight.

Findings

Pressures, Burdens, and Constraints on Bloggers

Routines eventuate, in part, from social, political or economic constraints and pressures, and findings reveal a number of these. Whereas the interviewed bloggers reported having few acute financial pressures and needs (using a server costs one blogger $60 per year), bloggers said they sought to increase the size of their audience or at least to maintain current readership. More than one blogger said a key turning point in the way they practice blogging was the moment they felt the gaze of the public eye. Realization that people are paying attention, through increased comments, hits and emails, has led these bloggers to adopt a more careful, dispassionate approach and tone. The national blogger said the increased readership made her and her co-author feel "a lot more responsible." The city blogger said the perception that people pay attention has led to less opinionating and more reporting and thoughtful analysis: "When I started to do more than just put my own thoughts up I put more into it, and when you do that and see what goes up and realize people are reading it and using it you worry about it more. I was more creative when I started, now I'm more deliberate." Similarly, the state blogger said in the beginning his blog was simply a rant:

> But then people started to read it, and I started trying to be more professional.... Now I'm getting about 1,000 readers a day.... Once I got to 100 readers I started to get more organized and started to take more responsibility for what I posted. Then I started to restrict what I put up there...I started to elaborate more, to reason more.... If I just rant about my views, they won't come.

Because of the heat of the public gaze, all the bloggers said they feel they must keep posting. As the national blogger said, "If I do not post every day, I feel terrible. I know people are waiting and wondering. I know they are impatient." The state/national blogger said if they miss a day, they may lose readers: "I know I don't have to have a post up at 10 in the morning, but there had better be one there by four in the afternoon."

These bloggers typically do not blog full time, for a living. The national blogger said finding time to blog around her day job is a challenge, and the state/national blogger said her production schedule is dictated in part by the fact that her children's schools let out at 3 p.m. Lack of resources means most bloggers do not benefit from division of labor and therefore have to be more creative with their time. "I don't have a staff," says the state blogger. "If you are not just regurgitating news from somewhere else you have to find time to do it, you have to be organized."

Finally, bloggers have to deal with the burdens of abundance. Ready access to other sites makes source material potentially limitless, and bloggers have unlimited space, a situation that is somewhat unique among media producers. Whereas constraints necessitate routines, so does a lack of limits. To render decision-making manageable, limits must be set, and as will be seen, bloggers have developed routine practices that narrow down possibilities.

Routines Related to Content and Source Material

Ideas and stories for posts are potentially endless for bloggers, but they have time pressures, stemming from other responsibilities in their lives and from expectations from readers. So, to make the task of finding potential post topics manageable, bloggers routinely rely on traditional news media websites or news aggregation sites offering a menu of content relevant to them. As the city blogger said: "For sources, if the topic is already an issue, I'll see what some of the mainstream has done."

The state blogger said he starts his day with an RSS[1] feed: "I get all the state political news I can from newspapers around the state and that sort of helps me decide on what topic to choose." The national blogger also uses RSS: "To look at everything myself would be unmanageable...I use it to scan sites and topics quickly and then, if I want to, I can go deeper." The national/state blogger uses a state newspaper and its website to find state stories. She said she also relies on a political blogger in the state who has a research assistant: "That blog gives us a list of state news stories every morning, and I use it not only to find a topic but to follow the links and read about it." These sites allow bloggers to quickly choose topics that are in the news, that fit their profile and that they know mirror their standards.

In addition to the source-material strategies cited above, bloggers also specialize. Specialization allows for development of the expertise needed to give readers what they want and to allay bloggers' concerns about readers' demands. Bloggers cannot keep up with all issues, but the bloggers who specialize have to research deeply on their specialties. Specializing is also one way bloggers can find a niche, which helps the blogger locate a community of dedicated readers and find the blogger's place within the larger blogosphere. As the national blogger said, "We look for what we can give them that no one else does." The sites to which a blogger decides to link sends readers further cues about the specific nature of the content they can expect to find on the blog. The state blogger said he links to related blogs in his "peer group": "I [will link] to someone who has relevant stuff. [Even] if it looks like they may be encroaching on my territory I'll still link." Finding one's peer group offers advantages, such as reliable sources for information from related blogs that will be applicable to one's specialty. This routine is not unlike the "beat" routine in journalism, which ensures a steady, predictable stream of stories (Tuchman, 1978).

Bloggers pursue a number of practices that relate to specializing or that help them specialize. For example, sticking with a developing story over time is a practice bloggers can say benefits the reader, but it is also a practice that helps the blogger by providing a justification for narrowing the range of story selection. As the state/national blogger said, "I'd rather post two or three detailed posts a day than 20…once I start writing about a story I try to follow it all the way to its conclusion."

Another strategy is to ignore or hold information that may attract criticism—to "bite my tongue," as the city blogger says. As the city blogger said: "I've ducked a couple of issues recently…because I wanted to be better informed. I didn't want to be wrong and have people know I was wrong and hadn't done all the work I should so I just avoided the topic." Bloggers may pass on such stories altogether. The state/national blogger said she often keeps information out of local stories:

> I've written about a candidate I know and said simply that the candidate has been in office too long and should go. I know more, and there are rumors I could check out… but I chose just to say that about being there too long.

Most bloggers have no staff to conduct research. Therefore, they are careful not to make unfounded claims because they might have to defend them from criticism, which takes time and energy. The state/national blogger said she wished she could do more research but that she could not afford to blog for a living: "I choose what to do and what I have time for. I am selective because I have to be."

Routines and Work Schedule

Bloggers say that a failure to post mars the blog's reputation, as does a failure to post material that is up to the standard of the blog's readers. Like working journalists, bloggers cannot expect to produce acceptable quality at the last minute. So bloggers set aside certain times to write their posts. Each of the bloggers interviewed cited blogging schedules that accommodate their other daily responsibilities. The national blogger has a day job and said finding time is challenging: "I write morning[s], at lunchtime or in the evening…I try to get in at least a post a day, ideally at peak hours." The city blogger also says time is an obstacle, but that he is getting the routine down: "I don't want it to affect my work. I write really early in the morning, and when I get off work, and at lunchtime. When I work in the early morning it is kind of a release for me." The state blogger says it is crucial to develop a schedule if you're not just "regurgitating news from somewhere else." He says he usually writes before going to bed, but he arranges for the posts to go live automatically at different times the following day, in response to the schedules he believes his readers keep.

Routines and the Blogging Genre

It is not possible to thoroughly assess here whether blogs are adopting standardized formats, though previous research has detected convergence toward conventions of identification through first name only, and an emphasis on text over visuals (Herring et al., 2007). The state blogger suggested that widespread adoption of blogging software may be leading to standardization: "I don't really have a set style I could describe. My process is to let the coding take care of it. If I quote more than two sentences, it will automatically be coded as a block quote and the [browser] will check to make sure the grammar and spelling are accurate."

However, all respondents mentioned the importance of merging fact-based information with commentary and analysis. All use a format in which excerpts of news information are followed by commentary. The national blogger said, "I try to build a base of news, news that I know is correct, and then I add my viewpoint. But whatever my views, the news has to back it up." The city blogger says he tries to package original content with opinion: "After I have something to write, I go back and try to use the commentary to add to it, to say why I did it or what it might mean." The state blogger offers the following anecdote:

> I haven't changed my views since I started the blog. I'm just packaging them differently. For example, [the state's attorney general] is one of the few people in state politics that I genuinely don't like, and I want him defeated in the election. When I started the blog, I'd just insult him, call him a moron, that sort of thing. Now I'll do my research and say what he's doing and say this is what he's wrong about, and this is why he's wrong.

Audience-Related Routines

Bloggers work to develop a niche in persona or style. Like a content specialization, a style niche helps bloggers find their place in the blogosphere and helps distinguish and define a blog for readers. A style niche is a kind of "media logic" or shared understanding between message producers and audiences about a media's format. For example, a detached, no-nonsense style of writing connotes "hard news" (Altheide & Snow, 1979; McQuail, 2005). Understanding and embracing a media logic allows bloggers to typify the sorts of sites they link to, making it easier to identify blogs that differ in style and then weed out links to these blogs, even if they are similar in content. For example, a blog with an edgy, light tone might not link to a blog with a heavy, academic style, even if the content specializations were similar. The city blogger says familiarity with a genre develops between the blogger and reader: "Eventually you start using a sort of shorthand." The blog persona calls to mind Goffman's (1959) "face work" concept, in which the projected characteristics of an individual's public "front" is shorthand to understanding an individual.

All the bloggers interviewed discussed the importance of establishing a persona. The national blogger was particularly intent on setting her blog apart: "I would take something that is a hot topic, as Barack Obama is now, for example. Then I wonder how I can have a different view...and not just be another voice saying the same thing as other people. I try to have an edge, to differentiate myself from other blogs." The city blogger discusses the importance of "positioning" oneself in the public eye so that the blogger is partly revealed and partly obscured: "It's interesting letting them know who I am and what I do—but not all of it." Under the public eye, he is more careful to post accurate statements, and says he would not "want people to think I do it for money." The state blogger said when his audience began to grow noticeably, he "started trying to be more professional." Now he says he nourishes a serious, newsy persona: "I take what I do very seriously...I am the top independent political blog in [the state] and I want to stay that way."

Like other media messengers, bloggers do not meet with their "client-audience." They must therefore develop conventions for constructing audiences, as is the case for other media producers (Ettema & Whitney, 1994; McQuail, 1997). For bloggers, the ability to instantly and quantitatively measure audience activity has a powerful influence on their constructs of the audience, an influence that has been observed in studies of mainstream online media as well (MacGregor, 2007). As the national blogger says, "We know our readers, we know our hits—we follow our traffic very closely." These statistics tell bloggers how many people are visiting their pages, what percent of these are new visitors, which stories are most visited, and what times during the day visits are more likely. The state blogger describes these statistics:

> I know that I get my first rush of readers between 8 to 10 in the morning, and then there are some people in the middle of the day and another rush between 5 to 7 at

night.... By looking at the hits and page views and unique visitors I think that most of my readers keep coming back—in other words, the people that come in the morning... come back during the day.

If one construct of the audience is the histogram, another is the "peer group." Each blog has its loyal followers who post comments and send emails, and bloggers tend to write to these individuals, much as journalists write for each other (McQuail, 1997, 2005). The state blogger says he has consulted his readers about endorsing candidates: "I asked [the people who come to my blog] if it would be presumptuous if I endorsed candidates and everyone said go ahead."

Consequences of the Routinization of Blogging

Findings show that bloggers adopt a set of routines to deal with their unique set of constraints, burdens, and pressures. Bloggers have the burdens/blessings of access to a deep reservoir of content and access to unlimited space within which to publish. They also have the mixed blessing of quantifiable instant feedback. Yet they are terribly short on time and do not have the luxury of dividing labor. In addition, they appear to be highly susceptible to pressure from the public gaze. All media producers experience this pressure to some degree, but media organizations have the financial and structural support to weather dips in approval. There is no real barrier to entry in the blogging world, and they are, therefore, more vulnerable to the fickleness of audiences. The counter to this argument is that the consequences are not material, and that bloggers stand to lose little that is concrete. Nevertheless, findings reveal that the consequences of losing status and losing face with readers feel very real to these bloggers.

How might the routines unearthed here shape blogging content? It is an important question, given that routines tend to serve the political, economic and social status quo. The routine of tracking news sites and easy availability of automated feeds suggests bloggers have little incentive to do actual reporting. Interestingly, the one blogger who did occasionally go out into the neighborhood and report was the least technically sophisticated in his blogging (the city blogger). He did not use a server and did not use automated feeds. And as the format for blog postings—i.e., the link to another site coupled with commentary—becomes solidified as the media logic for the blogging genre, it seems even less likely bloggers will pursue original reporting, as such posts would not fit well with the common understanding of the prevailing format. The routines of specialization and "ducking stories" also make it more probable that bloggers will find topic areas based on easily available information. And ducking stories is itself a way to dodge reporting.

The tendency to specialize also may discourage content that takes a big-picture view. Issues and topics are interrelated, and narrow analyses that leave out contexts and tangential connections may not always serve readers. Of course, bloggers may link to other sites that provide related information, even if they do not provide it themselves. As long as bloggers who post on similar content areas keep up with each others' work, a "big-picture view" may be available through a network of similar blogs.

The posturing involved in the routine of the persona niche, while helpful as a way of solidifying media logic with capricious readers, may lead bloggers away from substance and toward style, which itself can become a sort of straight-jacket. Writing with a certain tone—sarcasm, irony, "edge"—constrains the way meaning can be conveyed, particularly when the persona becomes concretized because one's readers grow to expect the persona. News organizations themselves are becoming more conscious of a perceived need to "brand" the product in order to identify with particular audiences, but they find this much more difficult to do because of the wider span of their audience. It would be interesting to see if bloggers with smaller audiences feel the need to develop specified personas more acutely than do bloggers with larger audiences. This question is particularly relevant given that many bloggers express an interest in acquiring more incoming links and growing audience.

Availability of instant, quantifiable audience feedback has an impact on decision making by bloggers in this study. This information may focus bloggers on the ephemeral preferences of readers, causing them to favor postings that already have high visibility and which are likely to gain short-term attention at the expense of less "sexy," more complex issues (MacGregor, 2007).

Both journalism and blogging have developed routines for maintaining audiences, but the routines differ because of unique burdens (e.g., blogging's unlimited resources and traditional organizational journalism's need for massive audiences) and unique advantages (bloggers' ability to duck content with which they are uncomfortable and journalists' division of labor). The routines of this young media genre are only now taking shape, and changes in these routines and their consequences bear watching, particularly as some blogs gain audience, revenue and workload. Such gains may lead to organizational forms of production because of a need to differentiate tasks, or they may lead to acquisition by existing media organizations, an increasingly common occurrence (Domingo & Heinonen, 2008). It would be most helpful if future study focused on the dynamics of changing forms and the changing relationship between journalism and blogging practices, as well as their causes and consequences.

NOTES

1. RSS (Really Simple Sindication) is a technology that allows to monitor the latest updates of weblogs and news sites without having to visit them. The headlines are sent to the RSS reader software or online RSS service chosen by the individual.

EPILOGUE

Toward a Sociology of Online News

Mark Deuze

In December 2002, the *Kidon Media-Link* international database of online professional news sources contained 14,111 organizations.[1] In December 2006, the same database listed 19,029 links to online news media.[2] After a "first wave" (Pryor, 2002) of electronic publishing experiments (1982–1992), a reluctant shift of news operations to online (1992–2002), and the gradual emergence of wireless broadband multimedia journalism (2002–onward), it is safe to say some kind of online, multimedia or cross-media, convergent, and otherwise distinctly digital journalism has become a structural part of the news media enterprise. This process has not been smooth, and the place of "online" in the organization of news work is far from ideal. The role and definition of journalism online is contested to say the least: a source of frustration for many, a platform for feverish techno-fetishist utopians and dystopians, a stepping stone toward newsroom restructuring for the struggling news business, a vehicle for underdetermined claims making by academic theorists and industry pundits alike. It is also a laboratory for experiments in workforce flexibility and labor exploitation in the news industry, as the work in online news can be done from anywhere (given the availability of the necessary hardware, software, and connectivity). Yet online journalism can also be the stomping ground for exciting, creative and innovative forms of news, where the conventions, roles and cultures of different types of media as well as producers and consumers of information converge. In short, online or digital journalism is here to stay, but its identity, place, and status are still very much "under construction"—as indeed anything online is.

The literature in the field of journalism studies is largely informed by the standards of research, education, routines, rituals and practices set by print journalism. This is remarkable and regrettable. Such a "print bias" in research and teaching journalism makes almost all theories of journalism intrinsically problematic when applied across the professional field as a whole. Studies looking at online journalism have emerged since the mid-1990s, as the Internet made its way into newsrooms and became both a reporting tool as well as a platform for news dissemination. Surveys among journalists in several countries show convincingly that a vast majority of journalists are now using the Internet regularly in their daily work. Several scholars have studied the effects of this process including the practices of Computer-Assisted Reporting (CAR), which is defined here as using Internet as a reporting tool. However, most of these studies again focused on print media adding online divisions, and on newspapers publishing their news (and archives) on websites. At the same time, an exciting new area of scholarly work—spanning Internet research, cybercultural studies, new media theory, and so on—has emerged, using the rich potential of the digital universe to gather, mine, and make sense of data in a much more connected, interactive and translocal field.

My point with both of these brief explorations in the fields of online journalism and new media studies is, of course, the same. To a large extent what happens at the boundaries or frontiers of a well-established social system, professional ideology or paradigm reveals more about what it reproduces than about what it purports to change. The study of online journalism is in fact an observation of how it is a rather distinctive form of journalism—both in the eyes of its practitioners and given the particular characteristics of its medium (Deuze & Dimoudi, 2002)—that is embedded in a broader news culture and occupational ideology (Deuze, 2005). As it emerges and grows, it is therefore expected to "fit" within broader and similar organizational forms of doing professional journalism, while it at the same time establishes some kind of particular identity and agency in the context of the larger system and institution it is part of.

Online News as Mini-Culture

As online newsrooms traditionally have been organized separately from their parent institutions and tend to be populated mostly by newcomers and contingent employees (Deuze, Neuberger, & Paulussen, 2004), these departments grew into their own "mini-cultures" (Ogbonna & Harris, 2006). These online journalism units have developed since the mid-1990s and are quite distinct, to some extent even countercultural departments within the profession, and their values, practices and ideals are found to be gradually changing the cultural mosaics of established news organizations (Deuze, 2003; Boczkowski, 2004a). Therefore, the sociology of online news always has to look at two processes at the same time. First, how specific news organizations and journalism as social institutions reproduce when adapting (iteratively, episodically) to disruptive technological and cultural influences. Secondly, how professionals, newsrooms, and other elements of the journalistic system, in fact, make change happen, creating new practices and ways of doing things out of components of the wider news culture.[3]

As a distinct professional practice—a fourth type of journalism—online journalism should be seen as journalism produced more or less exclusively for the World Wide Web. Online journalism has been functionally differentiated from other kinds of journalism by using its technological component as a determining factor in terms of a (operational as well as normative-ideal) definition—just like the fields of print, radio, and television journalism before it. The online journalist has to make decisions on which media format or formats best tell a certain story (multimediality), has to consider options for the public to respond, interact or even customize certain stories (interactivity), and thinks about ways to connect the story to other stories, archives, resources, and so on through hyperlinks (hypertextuality).

A second note has to be made regarding multimedia (Deuze, 2004) or convergent journalism (Quinn, 2005) as a practice or genre in the profession. Although one can consider multimedia journalism as an emerging distinct practice, the ongoing convergence (through digitization) of media modalities suggests that in one way or another, sooner or later, all journalism will have a multimedia component or defining characteristic, in that it will be possible to gather, edit, and deliver the news through all kinds of platforms using the same digital language of zeros and ones. Pragmatically this means that online, digital, and multimedia journalism all refer to the same professional development in the context of this particular chapter and book. A final note can be made about a third element constituting the practice of online news: "Web 2.0" and so-called "citizen media"; in various ways collaborative and participatory forms of journalism found particularly online, where audience members and professional journalists work together to produce news (Deuze, Bruns, & Neuberger, 2007).

Online News as Production of Culture

Digital journalism is constituted out of different iterations of online, multimedia, and/or participatory news production. Online news is in the broadest sense an instance of cultural production. The production of culture has been a traditional object of sociological investigation, which in the field of journalism studies has been spearheaded by the work of Gaye Tuchman (1978). "The production of culture perspective focuses on how the symbolic elements of culture are shaped by the systems within which they are created, distributed, evaluated, taught, and preserved" (Peterson & Anand, 2004, p. 311). It offers a specific framework for studying and analyzing the ways in which culture is produced, both within and across organizations, disciplines, professional groups, or even specific individualized contexts. Instead of assuming that whatever people do within a given social structure mirrors or reproduces that structure—which in journalism studies gets applied through a focus on news-room socialization and professional homogenization for example—the produc-tion of culture perspective "views both culture and social structure as elements in an ever-changing patchwork" (Peterson & Anand, 2004, p. 312). Such a focus on the often subtle, but important, processes of, at times, disruptive change is useful when trying to make sense of the emergence of this new form of journalism: the production of online news.

The creative process of work in the media industries in general, and online news in particular, is a fascinating object of study as the production of culture is in itself a cultural process. This means that neither the individual, nor the big corporation completely controls the production of culture—elements of social structure (the organization of work, the parameters set by time, budget and space, media ownership, and so on) and the norms, values, and ways of doing things of the professionals involved mutually influence and sometimes deter-mine each other. Indeed, cultural and economic concerns are not necessarily different, but in the context of media work rather must be seen as what Liz McFall (2004) calls "constituent material practice" (p.18): the combination of specific technical and organizational arrangements as these influence and are shaped by the generally idiosyncratic habits of individual media practitioners. An emphasis on the individualistic and idiosyncratic nature of media workers suggests that contemporary trends such as workforce casualization, technologi-cal and cultural convergence, and flexible productivity not only mean different things to different people, but are also differently articulated in the context of specific media products, genres, and organizations because of the ways in which departments, teams and individuals work together.

The production of culture perspective organizes its material (data, observa-tions, subjects) into a six-facet model of the production nexus, including technologies, law and regulation, industry structure, organization structure, occupational careers, and markets (Peterson & Anand, 2004). In general, media work can be seen as a particular instance of the production of culture as it takes place both within and outside of institutions, by both professionals and amateurs, both within and across particular media. In the following sections, I offer a review of the production of culture in the domain of online news, using the material of the research chapters of this book as my point of departure. Let me repeat that this approach underscores the argument that the social process among people and institutions through which cultural goods and services—such as (online) news—are produced influences their content. The production of culture perspective also makes it possible to draw comparisons across the diverse sites of culture creation (Deuze, 2007). A sociology of online news as a more or less distinct form of cultural production locates the practices and organization of journalism online within the wider context of professional journalism, yet articulates it with trends that are particular to its components.

A final note must be made about the method advocated and used throughout this volume: the ethnographic method. It is a crucial part of the toolkit for studying the production of culture, as it allows one to more closely observe the primary sociological object: the constantly changing, distinctly situational and renegotiated relationships between individuals and society, or rather, between agency and structure. The solution to the recurring dilemma of institutional structure and individual agency in the production of culture lies in understanding media conglomerates, organizations, and even temporary project-based collaborative networks as "inhabited institutions," where people do things together and in doing so continuously struggle for symbolic power within their respective fields of work. Tim Hallett and Mark Ventresca (2006) describe this inhabited institutions approach as on the one hand accepting that institutions provide the raw materials and guidelines for social interactions, and on the other hand, that the meanings of institutions and the work taking place in (and for) institutions are constructed and propelled forward by social interactions. Organizations, whether seen as enterprising individuals or multinational corporations, are dynamic entities, within which people constantly move in and out of their roles as consumers and producers, and where the production of culture is built on the particular work styles of media workers.

In what follows, I discuss online news in terms of the role technologies, law and regulation, industry and organizational structure, occupational careers, and markets play in how the emergence of digital journalism as a distinct professional practice gets meaning to the practitioners involved.

Technologies

Technology plays a major role in the production of news—always has. Particularly in the field of online news, technologies—hardware and software—not only act as trend amplifiers, but also as artifacts used by practitioners to establish their own professional identity. The research in this book (see especially the findings by Quandt, Puijk, Brannon, and Domingo in Part 2) suggests that the same technologies are used differently in different departments or areas of news work (e.g., print, online newsrooms), signaling how specific constellations of people and practices co-determine the appropriation of technology as a means to reproduce their own difference. Technologies furthermore act in news organizations to facilitate "collaboration without community": journalists in different departments to some extent work together through content management systems (CMS) and company intranets, but never meet in person. This, for example, allows media organizations to keep them separate—in different rooms, buildings or even countries as news organizations increasingly outsource editorial departments offshore (WAN, 2006).

The separation of different departments—however cooperative or interdependent—is furthermore supercharged by the shift toward an almost exclusively electronic news net, as reported by the various ethnographers in this volume. Online news professionals do practically all their work at their desks, using their connections to the wired world as the primary source of all things. As this also seems to be increasingly the case in other areas of news production, a picture emerges of an atomized profession, isolated and connected at the same time, yet also blind to each other (and thus itself), and the wider society it operates in.

On the other hand, the various observations report how technologies are not "cold machines," but often serve to allow practitioners or newsrooms to do more of the things they are already familiar with. Technologies are also acting as sources of conflict, tensions, and frustrations in news organizations, as their role is often imposed (from above, by management), they can at times disrupt existing informal hierarchies, and are generally poorly understood by the journalists involved and thereby giving more symbolic power to specialists such as software engineers and IT staffers.

At the heart of all these discussions about the role of technologies in online news is the central role a company's CMS plays. As every company or chain has its own often proprietary and always customized CMS, this piece of pricey software pretty much acts as a lightning rod for most or all of the issues involving imitation and change in newsroom cultures and practices. This is a fruitful area of investigation of technologies that enable and constrain at the same time, of technologies that truly represent the utopian ideals of cross-platform publishing while in their diverging appropriations and uses resembling organizational inertia and resistance.

Law, Policy, and Regulation

The immediacy of the online publishing environment poses some fascinating dilemmas in digital journalism. Such debates in terms of law and regulation are often framed in traditional contexts of libel (by allowing anyone to comment, blog or upload information to a news site) and intellectual property rights (by repurposing offline content produced by freelancers online for example). Despite the facts that most authors in this book do not directly address the legal challenges of digital journalism (see chapter 8 for the case of China), their findings illustrate how the emerging working routines in online news challenge or weaken the established legitimacy of journalism.

First, it seems that the policies regarding technologies and practices in news organizations are particular to different tasks, departments and individual people. Access to any or all areas of a company's CMS tends to be organized that way, indicating (and attributing) a different status to each member of such an organization. An informal hierarchy—between a central newsdesk and the different beats for example—thus gets reproduced and indeed formalized through network infrastructures, as determined by a company's policies.

Second, a specific element of online news work is the production of constantly changing, or "liquid" news. By introducing continuous news updates, breaking news segments, updates, and edits of developing news stories to the website of the media organization, its news output becomes less than stable. News online changes all the time. Stories are edited, tweaked, and changed—generally without explanation, and without explicit editorial policies guiding the users and producers. This not only contributes (and amplifies) the immediacy of online news, it changes specific notions of what a "story" is, or how a "deadline" functions, or when the news is actually a "finished" product (that would then be subsequent to the aforementioned traditional notions of law and regulations).

On a final note regarding the complex issues of law and policy one could look at the challenges a notion of liquid online news poses to the question of who is a journalist in the strictest sense of the word—especially now that an army of "semi-reporters" contribute to the ongoing reporting online, including but not limited to: IT people ("techies"), interns, freelancers, consumers (by posting comments with stories), and citizens (by blogging or uploading their own news and opinions at professional news websites). Most of these people do not enjoy the same protections and privileges professional journalists have, yet undeniably act ("work") as reporters and editors online, as Bruns, Lagerkvist, and Lowrey and Latta document in their chapters.

Industry and Organization Structures

It is important to note that online news and digital journalism consist of genres, practices, values, idea(l)s, and people that are very much part of a "work in progress," a constantly evolving set of cultural and structural determinants. It is therefore not surprising that the chapters in this volume articulate online news production with a sense of dynamism, perceptions of change, evolution, and revolution.

One of the key indicators of digital journalism as an emerging field of work is the distinct correlation of the organization of online news within the larger journalistic industry structure with ongoing debates in the profession about authority and legitimacy, as thoroughly explored by Cawley, García, and Colson and Heinderyckx in this book. Despite decade-long debates about the necessity of newsroom integration and convergence, generally speaking, online news departments still very much operate independent or separated from their offline counterparts virtually or (often: and) physically elsewhere. At the same time, their work is contested by their peers in the traditional newsroom, and some-times even by online journalists themselves as, to some extent, their workspaces are still very anarchic, lacking central oversight. Such a lack of managerial intervention indicates a lower status, with online staffers populating a perpetual in-between status: employed by a prestigious news brand, yet not acknowledged as fully-fledged members of the profession. In a way, online journalists undergo the typical migrant experience: not part of their "home country" anymore, but also never fully accepted by their "host country" either. Just as their news is liquid, they have to come to terms with a distinctly liquid, as in: unfinished, professional identity.

Within the messy, open, and relatively "unmanaged" workspaces of online news as documented in this book, the pace of work is accelerating. The production of online news is associated by its practitioners with speed (see especially chapter 7). At the same time—and perhaps because of this—the structure and organization of work in online news gets emphasized by its practitioners as being a work-in-progress, as dynamic and evolving. If we set such comments against the observation by most authors in this book that there seems to be little or no integration, collaboration or even contact between (members of) the online news team and the journalistic colleagues elsewhere in the media company, it certainly seems that the basic conditions for exploring mutual possibilities, knowledge-sharing, or indeed any other kind of professional "bonding" are not met.

Occupational Careers

The ethnographies in this book document a group of people that can be characterized as generally quite young, working in a stressful yet relatively undisturbed environment (with messy desks, an informal dress code, little or no oversight, and flat hierarchies). The work of online news staffers tends to be flexible and involves quite a bit of multiskilling: mastering different technologies and skills at the same time. Their relationship with the newsrooms of the mother-medium is at a certain distance, which means their careers develop rather autonomously. However, at the same time we see how these online reporters and editors adopt or mimic traditional journalistic values and practices in order to gain status and recognition with their offline colleagues.

The remarkable yet "under observed" work of online news staffers is furthermore complicated by the fact that many, if not most, digital journalists do not enjoy permanent salaried employment. These newsworkers also realize that their jobs are the first to go when the news company they are part of is suffering financially, as Cawley notes when reconstructing the recent history of *Ireland.com*. This kind of contingency in media work is dramatically invisible to colleagues elsewhere because of the aforementioned lack of integration between departments—leading journalists in the broadcast or print newsrooms to be (or feel) rather detached from their online counterparts. Indeed, in the occupational hierarchy, online journalists tend to be ranked quite low.

Markets and Audiences

Part of the motivation of news companies to have an online news division at all, or so it seems from the observations in this book, is to prepare for or counter increased competition in the journalistic marketplace. This is a negative ration-ale: investments in online are generally contingent and premised on perceived threats, rather than on a distinct vision or idea(l) about new or innovative forms of journalism.

At the same time, the various online journalists interviewed and observed by the authors in this book are developing their own rituals, best practices, skillsets and norms regarding the characteristics of their medium, particularly when it comes to addressing the online user. The consensus seems to be that online markets and audiences behave differently than those of traditional media, that their expectations of news are quite distinct, even that people understand (or "read") news differently online. Thus, within a context of a precarious work environment and low investments in truly innovative projects, digital journalists generally seem to be developing their own set of practices and values that keep the newly interactive or empowered audience at bay, outside. In a way this new construction of the audience online—as active, interactive, connected, inter-ested—serves to validate the work that online news staffers have to do (or are doing) beyond their limited job descriptions, outside of the view of their colleagues elsewhere, but in many ways well within fairly traditional parameters of what it means to be a "professional" journalist.

Discussion

Two different stories about online news emerge from the work in this book. On the one hand, one of a struggling group of low-status employees of news companies that cannot or will not see the online divisions as part of their core product and workflow. At best, working in the online newsrooms thus becomes like the jobs of game testers in videogames or below-the-line labor on the sets of film and television productions: something you do just to get "in," to hopefully get noticed as you try to make up your way into the dominant sectors of the profession. The second story is one of a scattered, informally disorgan-ized, yet motivated workforce that uses its limited resources and lack of formal power in the news organization to shape and cultivate its own professional identity, using technologies (CMS) and new perceptions of audiences to distinguish themselves and their work from colleagues elsewhere in journalism, while at the same rereading some fairly traditional boundaries between "them" (the media workers, professionals) and "us" (the audience). A sociology of online news thus has to take note of old and new processes of inclusion and exclusion (of people and ideas) in the ongoing critical debates about how the profession needs to change to adapt to new realities in the way people use media in their everyday lives.

For a research agenda in online news, I would advocate a more rigorous approach to articulating, studying, mapping, and explaining agency in news-work. It is clear that much of the work in online news is repetitive, and just reproduces the work that was done by others. However, one cannot ignore the emergence of numerous new formats, genres, and innovations in the production of online news. Considering the observations collected in this book, one wonders how these online staffers pull this off. It is in that particular situational context that research on agency (beyond simple imitation or change frameworks) should take place. Second, we have to explore the construction of audiences that is taking place online. Is the proposed idea of an interactive, participatory, and connected user a legitimate tool to inspire new ways of doing journalism, or does it just foster informal processes of articulating a distinct professional identity in online news? To what extent does such an "imagined audience" (Ang, 1991) contribute to new ways of excluding certain (groups of) citizens from the contemporary news agenda? Third, our research agenda should attempt more rigorously to connect issues of producer agency and user involvement with content. What gets produced in online news that truly pushes the boundaries of the profession of journalism, and how do we—as scholars—and people—as consumers—see it when this happens? Are the rare examples of truly innovative or experimental forms of journalism online brought to us by a online news staff working hard with limited resources to provide society with better information, or is it more likely that these are expressions of an emerging professional group trying to be remarkable—not so much for their audiences, but rather toward peers?

NOTES

1. The Kidon Media-Link database listed on December 6, 2002: 10,721 newspapers, 239 agencies, 1,004 Internet-only news publications, 273 magazines, 788 radio stations, 1,065 television stations, and 22 teletext services online (Source: webmaster Kees van der Griendt in personal email communication, December 6, 2002).

2. See www.kidon.com/media-link

3. The argument in this paragraph integrates insights from neo-institutionalism, social system approaches, and organizational theories; for an overview of contemporary theories in journalism research see Löffelholz and Weaver (2007).

References

Abbate, J. (1999). *Inventing the Internet*. Cambridge, MA: MIT Press.

Akrich, M., & Latour, B. (1992). The de-scription of technical objects. In W.E. Bijker & J. Law (Eds.), *Shaping technology/building society* (pp. 205–224). Cambridge, MA: MIT Press.

Allen, S. (1994). What is media anthropology? A personal view and a suggested structure. In S. Allen (Ed.), *Media anthropology: Informing global citizens* (pp. 15–32). Westport, CT: Bergin & Garvey.

Altheide, D. (1976). *Creating reality: How TV news distorts events*. Beverly Hills, CA: Sage.

Altheide, D., & Snow, R. (1979). *Media logic*. Beverly Hill, CA: Sage.

Altmeppen, K.-D. (1999). *Redaktionen als Koordinationszentren: Beobachtunjournalistischen Handelns* [Newsrooms as coordination centres: Observations of journalistic action]. Opladen, Germany: VS Verlag.

Altmeppen, K.-D., Donges, P., & Engels, K. (1999). *Transformation im Journalismus. Journalistische Qualifikationen im privaten Rundfunk am Beispiel norddeutscher Sender* [Transformation in journalism. Journalistic qualifications in private broadcasting. The example of broadcasters in northern Germany]. Berlin: Vistas.

Ang, I. (1991). *Desperately seeking the audience*. London: Routledge.

Atkinson, P., & Hammersley, M. (1994). Ethnography and participant observation. In N.K. Denzin & Y.S. Lincoln (Eds.), *Handbook of qualitative research* (pp. 248–261). Thousand Oaks, CA: Sage.

Atton, C., & Wickenden, E. (2005). Sourcing routines and representation in alternative journalism: A case study approach. *Journalism Studies, 6*(3), 347–359.

Baker & McKenzie Ltd. (2001). *China and the Internet: Essential legislation*. Hong Kong: Asia Information Associates.

Barabasi, A.L., & Albert, R. (1999). Emergence of scaling in random networks. *Science, 286*(5439), 509–512.

Baszanger, I., & Dodier, N. (2004). Ethnography: Relating the part to the whole. In D. Silverman (Ed.), *Qualitative research: Theory, method and practice* (pp. 9–34). London: Sage (2nd ed).

Baum, J.A.C. (2001). Organizational ecology. In S.R. Clegg, C. Hardy, & W.R. Nord (Eds.), *Handbook of organization studies*. London: Sage.

Becker, H. (1982). *Art worlds*. Beverly Hills, CA: Sage.

Becker, L.B., Vlad, T., & McLean, J. (2007). *Annual survey of journalism & mass communication graduates*. The Cox Center, Grady College, University of Georgia. Retrieved from: http://www.grady.uga.edu/ANNUALSURVEYS/grd06/grdrpt2006_merged_full_v6.pdf

Benker, S. (2001). Kreativer Computerfreak [Creative computer freak]. *Journalist, 10*, 61–63.

Benkler, Y. (2006). *The wealth of networks: How social production transforms markets and freedom*. New Haven, CT: Yale University Press.

Berger, P.L., & Luckman, T. (1966). *The social construction of reality: A treatise in the sociology of knowledge*. Garden City, NY: Anchor Books.

Bijker, W.E. (1995). Sociohistorical technology studies. In S. Jasanoff, G.E. Markle, J.C. Petersen, & T. Pinch (Eds.), *Handbook of science and technology studies* (pp. 229–256). London: Sage.

Bijker, W.E., & Pinch, T. (1987). The social construction of facts and artifacts: Or how the sociology of science and the sociology of technology might benefit each other. In W.E. Bijker, T.P. Hughes, & T. Pinch, *The social construction of technological systems*. Cambridge, MA: MIT Press.

Boczkowski, P.J. (1999). Understanding the development of online newspapers. *New Media & Society, 1*(1), 101–126.

———. (2002). The development and use of online newspapers: What research tells us and what we might want to know. In L.A. Lievrouw & S. Livingstone (Eds.), *Handbook of new media* (pp. 270–286). London: Sage.

——— .(2004a). *Digitizing the news: Innovation in online newspapers*. Cambridge, MA: MIT Press.

———. (2004b). The processes of adopting multimedia and interactivity in three online newsrooms. *Journal of Communication, 54*(2), 197–213.

———. (2004c). Books to think with. *New Media & Society, 6*(1), 144–150.

Boczkowski, P.J., & Lievrouw, L.A. (2007). Bridging communication studies and science and technology studies: Scholarship on media and information technologies. In E.J. Hackett, O. Amsterdamska, M. Lynch, & J. Wajcman (Eds.), *The handbook of science and technology studies*. Cambridge, MA: MIT Press (3rd ed.).

Bowman, S., & Willis, C. (2003). *We media: How audiences are shaping the future of news and information*. Reston, VA: The Media Center at the American Press Institute. Retrieved from: http://www.hypergene.net/wemedia/download/we_media.pdf

Brady, A.-M. (2007). *Marketing dictatorship: Propaganda and thought work in China*. Buffalo, NY: Rowman & Littlefield.

Brannon, J. (1999). *Maximizing the medium: Assessing impediments to performing multimedia journalism at three news web sites*. Doctoral dissertation, University of Maryland.

Bressers, B. (2006). Promise and reality: The integration of print and online versions of major metropolitan newspapers. *The International Journal on Media Management, 8*(3), 134–145.

Bressers, B., & Meeds, R. (2007). Newspapers and their online editions: Factors that influence successful integration. *Web Journal of Mass Communication Research, 10*. Retrieved from: http://www.scripps.ohiou.edu/wjmcr/vol10/

Brewer, J.D. (2000). *Ethnography*. Buckingham, U.K.: Open University Press.

Brügger, N. (2005). *Archiving websites: General considerations and strategies*. Aarhus, Denmark: Centre for Internet Research.

Bruns, A. (2005). *Gatewatching: Collaborative online news production*. New York: Peter Lang.

———. (2007a). Produsage: Towards a broader framework for user-led content creation. Paper presented at *Creativity & Cognition 6*, Washington, DC.

———. (2007b). The future is user-led: The path towards widespread produsage. Paper presented at *PerthDAC*, Perth, Australia.

———. (2007c). Methodologies for mapping the political blogosphere: An exploration using the issue crawler research tool. *First Monday, 12*(5). Retrieved from: http://firstmonday.org/issues/issue12_5/bruns/index.html

Buckalew, J.K. (1970). News elements and election by television news editors. *Journal of Broadcasting, 14*, 47–54.

Callon, M. (1987). Society in the making: The study of technology as a tool for sociological analysis. In W.E. Bijker, T.P. Hughes, & T. Pinch (Eds.), *The social construction of technological systems* (pp. 83–110). Cambridge, MA: MIT Press.

Carey, J. (2005). Historical pragmatism and the internet. *New Media & Society, 7*(4), 443–455.

Cassel, C., & Symon, G. (2004). *Essential guide to qualitative methods in organizational research*. London: Sage.

Castelló, E., & Domingo, D. (2004). Poking our noses into the production process: Benefits and risks of researchers' involvement in new media projects. Paper presented at the *2nd Symposium New Research for New Media: Innovative Research Methodologies*, Tarragona, Catalonia. Retrieved from: http://www.makingonlinenews.net/docs/castello_domingo_nrnm2004.pdf

Chan, A. (2002). From propaganda to hegemony: Jiaodian Fangtan and China's media policy. *Journal of Contemporary China, 11*(30), 35–52.

Chan, J.M., Lee, F.L.F., & Pan, Z. (2006). Online news meets established journalism: How China's journalists evaluate the credibility of news websites. *New Media & Society, 8*, 925–945.

Chung, D. (2003). *Toward interactivity: How news websites use interactive features and why it matters*. Doctoral dissertation, Indiana University.

———. (2007). Profits and perils: Online news producers' perceptions of interactivity and uses of interactive features. *Convergence, 13*(1), 43–61.

Colon, A. (2000, May/June). The multimedia newsroom. *Columbia Journalism Review*. Retrieved from: http://backissues.cjrarchives.org/year/00/2/colon.asp

Cottle, S. (2000). New(s) times: Towards a 'second wave' of news ethnography. *Communications, 25*(1), 19–41.

———. (Ed.). (2003). *Media organization and production*. London: Sage.

———. (2007). Ethnography and journalism: New(s) departures in the field. *Sociology Compass, 1*(1), 1–16.

Cottle, S., & Ashton, M. (1999). From BBC newsroom to BBC newscentre: On changing technology and journalist practices. *Convergence, 5*(3), 22–43.

Creswell, J. (1998). *Qualitative inquiry and research design: Choosing among five traditions.* Thousand Oaks, CA: Sage.

Dailey, L., Demo, L., & Spillman, M. (2005). The convergence continuum: A model for studying collaboration between media newsrooms. *Atlantic Journal of Communication, 13*(3), 150–168.

Dalberg, V. (2001). *Representasjon av Kontekst: Flermedial publisering på tvers av praksisfellesskap i NRK.* [Representation of Context: Multimedia publishing across praxis communities in NRK] Master's thesis, University of Oslo.

De Burgh, H. (2003). *The Chinese journalist: Mediating information in the world's most populous country.* London: Routledge.

Delamont, S. (2004). Ethnography and Participant Observation. In C. Seale, G. Gobo, J. Gubrium, & D. Silverman (Eds.), *Qualitative research practice* (pp. 217–229). London: Sage.

Delwiche, A. (2004). Agenda-setting, opinion leadership and the world of web logs. Paper presented at the annual conference of the *International Communication Association,* New Orleans, LA.

Demo, L., Spillman, M., & Dailey, L. (2004/2005). *Newsroom partnership survey executive summary* and *Television newsroom partnership survey executive summary.* Muncie, IN: Ball State University Center for Media Design. Retrieved from: http://web.bsu.edu/mediasurvey/

Denzin, N.K. (1997). *Interpretive Ethnography: Ethnographic practices for the 21st century.* Thousand Oaks, CA: Sage.

Deuze, M. (1998). The web-communicators: Issues in research into online journalism and journalists. *First Monday, 3*(12). Retrieved from: http://www.firstmonday.org/issues/issue3_12/deuze/

———. (1999). Journalism and the Web: An analysis of skills and standards in an online environment. *Gazette, 61*(5), 373–390.

———. (2001). Understanding the impact of the internet: On new media professionalism, mindsets and buzzwords. *Ejournalist, 1*(1). Retrieved from: http://www.ejournalism.au.com/ejournalist/deuze.pdf

———. (2003). The web and its journalisms: Considering the consequences of different types of news media online. *New Media & Society, 5*(2), 203–220.

———. (2004). What is multimedia journalism? *Journalism Studies, 5*(2), 139–152.

————. (2005). What is journalism? Professional identity and ideology of journalists reconsidered. *Journalism Theory Practice & Criticism, 6*(4), 443–465.

————. (2007). *Media work.* Cambridge, U.K.: Polity Press.

Deuze, M., Bruns, A., & Neuberger, C. (2007). Preparing for an age of participatory news. *Journalism Practice, 1*(4), 322–338.

Deuze, M., & Dimoudi, C. (2002). Online journalists in the Netherlands: Towards a profile of a new profession. *Journalism, 3*(1), 85–100.

Deuze, M., Neuberger, C., & Paulussen, S. (2004). Journalism education and online journalists in Belgium, Germany, and the Netherlands. *Journalism Studies, 5*(1), 19–29.

Deuze, M., & Paulussen, S. (2002). Online journalism in the low countries: Basic, occupational and professional characteristics of online journalists in Flanders and the Netherlands. *European Journal of Communication, 17,* 237–245.

Domingo, D. (2003). Ethnography for new media studies: A field report of its weaknesses and benefits. Paper presented at the *1ˢᵗ Symposium New Research for New Media: Innovative Research Methodologies,* Minneapolis, MN. Retrieved from: http://www.makingonlinenews.net/docs/domingo_nrnm2003.pdf

————. (2004). Comparing professional routines and values in online newsrooms: A reflection from a case study. Paper presented at the annual *IAMCR Conference,* Porto Alegre, Brazil. Retrieved from: http://www.makingonlinenews.net/docs/domingo_iamcr2004.pdf

————. (2005). The difficult shift from utopia to realism in the Internet era. A decade of online journalism research: theories, methodologies, results and challenges. Paper presented at the *First European Communication Conference,* Amsterdam, the Netherlands. Retrieved from: http://www.makingonlinenews.net/docs/domingo_amsterdam2005.pdf

————. (2006). *Inventing online journalism: Development of the Internet as a news medium in four Catalan newsrooms.* Doctoral dissertation. Universitat Autònoma de Barcelona, Bellaterra, Catalonia. Retrieved from: http://www.tesisenxarxa.net/TDX-1219106-153347

Domingo, D. et al. (2007). Four dimensions of journalistic convergence: A preliminary approach to current media trends at Spain. Paper presented at the *8ᵗʰ International Symposium on Online Journalism,* Austin, TX. Retrieved from: http://online.journalism.utexas.edu/2007/papers/Domingo.pdf

Domingo, D., & Heinonen, A. (2008). Weblogs and journalism: A typology to explore the blurring boundaries. *Nordicom Review, 29.*

Domingo, D., Quandt, T., Heinonen, A., Paulussen, S., Singer, J., & Vujnovic, M. (2008). Participatory journalism practices in the media and beyond: An international comparative study of initiatives in online newspapers. *Journalism Practice, 2(3).*

Duhe, S.F., Mortimer, M.M., & Chow, S.S. (2004). Convergence in North American TV newsrooms: A nationwide look. *Convergence, 10,* 81–104.

Dupagne, M., & Garrison, B. (2006). The meaning and influence of convergence: A qualitative case study of newsroom work at the Tampa News Center. *Journalism Studies, 7*(2), 237–255.

Dylko, I.B., & Kosicki, G. (2006). Sociology of news and new media. Paper presented at the annual meeting of the *AEJMC*, San Francisco, CA.

Eisenhart, D. (1994). *Publishing in the information age: A new management framework for the digital era.* Westport, CT: Quorum Books.

Eldridge, J. (Ed.). (1995). *Glasgow Media Group reader.* New York: Routledge.

Eliasoph, N. (1997). Routines and the making of oppositional news. In D. Berkowitz (Ed.), *Social meaning of news* (pp. 230–254). Thousand Oaks, CA: Sage.

Engebritsen, N. (2007). *Digitaliseringen: Vilken roll spelar den?* [Digitalisation: What role does it play?]. Master's thesis, Lillehammer University College, Norway.

Epstein, E. (1974). *News from nowhere.* New York: Vintage Books.

Erdal, I.J. (2007). Negotiating convergence. In T. Storsul & D. Stuedahl (Eds.), *Ambivalence towards convergence.* Göteborg, Sweden: Nordicom.

Eriksen, L.B., & Ihlström, C. (2000). Evolution of the web news genre: The slow move beyond the print metaphor. In *Proceedings of the 33rd Hawaii International Conference on System Sciences.* Los Alamitos, CA: IEEE.

Ettema, J.S., & Whitney, D.C. (Eds). (1994). *Audience-making: How the media create the audience.* Thousand Oaks, CA: Sage.

Ettema, J., Whitney, D.C., & Wackman, D. (1987). Professional mass communicators. In C.R. Berger and S.H. Chaffee (Eds.), *Handbook of Communication Science* (pp. 747–780). Newbury Park, CA: Sage.

Filak, V. (2004). Cultural convergence: Intergroup bias among journalists and its impact on convergence. *Atlantic Journal of Communication, 12*(4), 216–232.

Fishman, M. (1980). *Manufacturing the news.* Austin: University of Texas Press.

Flichy, P. (1999). The construction of new digital media. *New Media & Society, 1*(1), 33–41.

———. (2006). New media history. In L.A. Lievrouw & S. Livingstone (Eds.), *Handbook of New Media.* London: Sage (2nd ed.).

Fox, J., Glory K., & Volkan S. (2007). No joke: A comparison of substance in *The Daily Show with Jon Stewart* and broadcast network television coverage of the 2004 Presidential Campaign. *Journal of Broadcasting & Electronic Media, 51*(2), 213–227.

Friend, C., & Singer, J. (2007). *Online journalism ethics: Traditions and transitions.* Armonk, NY: M.E. Sharpe.

Gallo, J. (2004). Weblog journalism: Between infiltration and integration. In L.J. Gurak, S. Antonijevic, L. Johnson, C. Ratliff, & J. Reyman (Eds.), *Into the Blogosophere: Rhetoric, community, and culture of weblogs*. Retrieved from: http://blog.lib.umn.edu/blogosphere/ weblog_journalism.html

Gans, H. (1980). *Deciding what's news*. New York: Vintage Books.

García, E.P. (2004). Online newspapers and the labyrinths of access. Paper presented at the 2nd *Symposium New Research for New Media: Innovative research methodologies*, Tarragona, Catalonia. Retrieved from: http://www.makingonlinenews.net/docs/garcia_nrnm2004.pdf

Geertz, C. (1973). *Interpretation of cultures*. New York: Basic Books.

Giddens, A. (1984). *The constitution of society: Outline of the theory of structuration*. Cambridge, U.K.: Polity.

Gillmor, D. (2004). *We the media: Grassroots journalism by the people, for the people*. Cambridge, MA: O'Reilly.

Gitlin, T. (1980). *The whole world is watching: Mass media in the making and unmaking of the new left*. Berkeley: University of California Press.

Glaister, D. (2005, June 22). LA Times 'wikitorial' gives editors red faces. *The Guardian Unlimited*. Retrieved from: http://www.guardian.co.uk/technology/2005/jun/22/ media.pressandpublishing

Glaser, M. (2003) The infectious desire to be linked in the blogosphere. *Nieman Reports, 57*(3), 86–88. Retrieved from: http://www.nieman.harvard.edu/reports/03-3NRfall/V57N3.pdf

———. (2004, April 7). Lack of unions makes Florida the convergence state. *Online Journalism Review*. Retrieved from: http://www.ojr.org/ojr/glaser/1081317274.php

Goffman, E. (1959). *The presentation of self in everyday life*. Garden City, NY: Anchor Books.

Golder, S.A., & Huberman, B.A. (2005). The structure of collaborative tagging systems. Ithaca, NY: Cornell University. Retrieved from : http://arxiv.org/ftp/cs/papers/0508/0508082.pdf

Golding, P., & Elliot, P. (1979). *Making the news*. New York: Longman.

Görke, A. (2000). Systemtheorie weiterdenken. Das Denken in Systemen al Herausforderung für die Journalismusforschung [Thinking systems theory farther. Thinking in systems as a challenge to journalism research]. In M. Löffelholz (Ed.), *Theorien des Journalismus: Ein diskursives Handbuch* [Journalism theories: A discursive handbook] (pp. 435–454). Opladen, Germany: VS Verlag.

Goussous, D. (2007). *Is weblogging an effective tool to promote democracy and freedom of expression? Evidence from Egypt and Jordan*. Master's thesis, University of Leeds, UK.

Haas, T. (2005). From 'public journalism' to 'public's journalism'? Rhetoric and reality in the discourse on weblogs. *Journalism Studies, 6*(3), 387–396.

Hallett, T., & Ventresca, M. (2006). Inhabited institutions: Social interactions and organizational forms in Gouldner's patterns of industrial bureaucracy. *Theory and Society, 35*(2), 213–236.

Halloran, J.D. (1981). The context of mass communications research. In E.G. McAnany, J. Schnitman, & N. Janus (Eds.), *Communication and social structure: Critical studies in mass media research* (pp. 21–50). New York: Praeger.

Hammersley, M., & Atkinson, P. (1995). *Ethnography: Principles in practice*. London: Routledge (2nd ed.).

Hammond, S., Petersen, D., & Thomsen, S. (2000). Print, broadcast and online convergence in the newsroom. *Journalism & Mass Communication Educator, 55*(2), 16–26.

Hargittai, E. (2005). Cross-ideological conversations among bloggers. Paper presented at the annual conference of the *International Communication Association,* New York.

Harper, C. (2003). Journalism in a digital age. In H. Jenkins & D. Thorburn (Eds.), *Democracy and new media* (pp. 271–280). Cambridge, MA: MIT Press.

He, Q. (2004). Media control in China. *China Rights Forum*, 1. Retrieved from: http://www.hrichina.org/fs/view/downloadables/pdf/downloadable-resources/a1_MediaControl1.2004.pdf

He, Z., & Zhu, J. (2002). The ecology of online newspapers: The case of China. *Media, Culture & Society, 24*(1), 121–137.

Heinonen, A. (1999). *Journalism in the age of the net: Changing society, changing profession*. Doctoral dissertation. Tampere: Acta Universitatis Tamperensis. Retrieved from: http://acta.uta.fi/pdf/951-44-5349-2.pdf

Hermida, A., & Thurman, N. (2007). Comments please: How the British news media is struggling with user-generated content. Paper presented at the *8th International Symposium on Online Journalism,* Austin, TX. Retrieved from: http://online.journalism.utexas.edu/2007/papers/Hermida.pdf

Herring, S., Scheidt, L., Kouper, I., & Wright, E. (2007). Longitudinal content analysis of blogs, 2003–2004. In M. Tremayne (Ed.), *Blogging, citizenship and the future of media* (pp. 3–20). New York: Routledge.

Hine, C. (2000). *Virtual ethnography*. Thousand Oaks, CA: Sage.

———. (Ed.). (2005). *Virtual methods*. Oxford: Berg.

Huang, E., Davison, K., Shreve, S., Davis, T., Bettendorf, E., & Nair, A. (2006). Facing the challenges of convergence: Media professionals' concerns of working across media platforms. *Convergence, 12*(1), 83–98.

Huang, E., Rademakers, L., Fayemiwo, M., & Dunlap, L. (2004). Converged journalism and quality: A case study of *The Tampa Tribune* news stories. *Convergence, 10*(4), 73–91.

Hughes, T. (1994). Technological momentum. In M.R. Smith & L. Marx (Eds.), *Does technology drive history? The dilemma of technological determinism* (pp. 101–113). Cambridge, MA: MIT Press.

Huxford, J., & Duda, N. (2000). *Cultures in collision: Newspapers and the Internet*. Paper presented at the annual conference of the *International Communication Association*, Acapulco, Mexico.

Innis, H. (1991). *The bias of communication*. Toronto: University of Toronto Press (revised ed.).

Jankowski, N.W., & van Selm, M. (2000). Traditional news media online: an examination of added values. *Communications, 25*(1), 85–101.

————. (2005). Epilogue: Methodological concerns and innovations in Internet research. In C. Hine (Ed.), *Virtual methods: Issues in social research on the Internet*. Oxford: Berg.

Jenkins, H. (2001). Convergence ? I diverge. *Technology Review, 104*, 93.

————. (2006). *Convergence culture: Where old and new media collide*. New York: New York University Press.

Kansas, D., & Gitlin, T. (1999). What's the rush? *Media Studies Journal, 13*(2), 72–76.

Katz, M. (2005). Internet gurus: Sociology and ideology of prominent Internet professionals. *Trípodos*, Extra 2005, 157–170. Retrieved from: http://cicr.blanquerna.url.edu/2005/Abstracts/PDFsComunicacions/vol1/03/Katz_Merav.pdf

Kawamoto, K. (1998). News and information at the crossroads: Making sense of the new online environment in the context of traditional mass comunication study. In D.L. Borden & K. Harvey (Ed.), *The electronic grapevine: Rumor, reputation, and reporting in the new on-line environment*. Mahwah, NJ: Erlbaum.

Kenney, K., Gorelik, A., & Mwangi, S. (2000). Interactive features of online newspapers. *First Monday, 5*(1). Retrieved from: http://www.firstmonday.org/ISSUES/issue5_1/kenney/index.html

Ketterer, S., Weir, T., Smethers, J., & Back, J. (2004). Case study shows limited benefits of convergence. *Newspaper Research Journal, 25*(3), 52–65.

Kim, H., & Blood, R. (2005). Blogs as new media on the Internet. *Review of Communication, 5*(2/3), 100–108.

King, N. (1994). The qualitative research interview. In C. Cassell & G. Symon (Eds), *Qualitative methods in organizational research. A practical guide*. London: Sage

Kirchhoff, L., Bruns, A., & Nicolai, T. (2007). Investigating the impact of the blogosphere: Using PageRank to determine the distribution of attention. Paper presented at the *Association of Internet Researchers* conference, Vancouver, Canada.

Kopper, G., Kolthoff, A., & Czepek, A. (2000). Online journalism: A report on current and continuing research and major questions in the international discussion. *Journalism Studies, 2*(1), 499–512.

Kristof, N.D. (2001, April 22). Fruits of democracy; guess who's a Chinese journalist now? *New York Times*.

Kuhnke, M. (1998). *Online-Redaktionen deutscher Medienunternehmen* [Online newsrooms in German media companies]. Diploma thesis, Universität Hannover, Germany.

Lagerkvist, J. (2005). The rise of public opinion in the People's Republic of China. *China: An International Journal, 3*(1), 119–130.

————. (2006). *The internet in China: Unlocking and containing the public sphere.* Lund, Sweden: Lund University Press.

————. (2008). Internet ideotainment in the PRC: National responses to cultural globalization. *Journal of Contemporary China,17*(54), 121–140.

Lasica, J.D. (2003). Blogs and journalism need each other. *Nieman Reports, 57*(3), 70–74. Retrieved from: http://www.nieman.harvard.edu/reports/03-3NRfall/V57N3.pdf

Latham, K. (2000). Nothing but the truth: News media, power and hegemony in South China. *China Quarterly, 163,* 633–654.

Latour, B. (1993). Ethnography of a high-tech case. In P. Lemonnier (Ed.), *Technological choice. Transformations in material culture since the neolithic.* London, New York: Routledge.

————. (2005). *Reassembling the social: An introduction to actor-network theory.* New York: Oxford University Press.

Lawson-Borders, G. (2003). Integrating new media and old media: Seven observations of convergence as a strategy for best practices in media organizations. *The International Journal on Media Management, 5*(2), 91–99.

Lee, C.C. (1990). Massmedia: Of China, about China. In C.C. Lee (Ed.), *Voices of China: The interplay of politics and journalism in China.* New York: Guilford Press.

Lemonnier, P. (1993). *Technological choice: Transformations in material culture since the neolithic.* London: Routledge.

Lenhart, A., & Fox, S. (2006). Bloggers: A portrait of the internet's new storytellers. Washington, DC: Pew Internet and American Life Project. Retrieved from: http://www.pewinternet.org/PPF/r/186/report_display.asp

Liebling, A.J. (1960, May 14). Do you belong in journalism? *The New Yorker*, p. 109.

Liebman, B.L. (2005). Watchdog or demagogue: The media in the Chinese legal system. *Columbia Law Review, 105*(1). Retrieved from: http://columbialawreview.org/pdf/Liebman-Web.pdf

Lievrouw, L.A. (2002). Introduction to part two: Technology design and development. In L.A. Lievrouw, & S. Livingstone (Eds.), *Handbook of new media.* London: Sage.

————. (2004). What's changed about new media? *New Media & Society, 6*(1), 9–15.

————. (2006). New media design and development: Difussion of innovations v social shaping of technology. In L.A. Lievrouw & S. Livingstone (Eds.), *Handbook of new media: Social shaping and consequences of ICTs* (pp. 246–265). London: Sage (2nd ed.).

Lievrouw, L.A. et al. (2001). Bridging the subdisciplines: An overview of communication and technology research. *Communication Yearbook, 24,* 271–296.

Lindlof, T.R. (1995). *Qualitative communication research methods.* London: Sage.

Lindlof, T.R., & Taylor, B. (2002). *Qualitative communication research methods.* Thousand Oaks, CA: Sage (2nd ed.).

Löffelholz, M., & Weaver, D. (Eds.). (2007). *Global journalism research*. Malden, MA: Blackwell.

Logan, R. (1995). *The fifth language: Learning a living in the computer age*. Toronto: Stoddart.

Lowe G.F., & Hujanen, T. (Eds.). (2004). *Broadcasting & convergence: New articulations of the public service remit*. Göteborg, Sweden: Nordicom.

Lowe G.F., & Jauert, P. (Eds.). (2005). *Cultural dilemmas in public service Broadcasting*. Göteborg, Sweden: Nordicom.

Lowrey, W. (2006). Mapping the journalism-blogging relationship. *Journalism, 7*(4), 477–500.

Lowrey, W., & Anderson, W. (2005). The journalist behind the curtain: Participatory functions on the internet and their impact on perceptions of the work of Journalism. *Journal of Computer-Mediated Comunication, 10*(3). Retrieved from: http://jcmc.indiana.edu/vol10/issue3/lowrey.html

Lowrey, W., & Mackay, J. (2006). Journalism and blogging: A test of a model of occupational competition. Paper presented at the annual meeting of the *AEJMC,* San Francisco, CA.

Lynch, D. (1999). *After the propaganda state: Media, politics, and 'thought work' in reformed China*. Stanford, CA: Stanford University Press.

MacGregor, P. (2007). Tracking the online audience: Metric data start a subtle revolution. *Journalism Studies, 8*(2), 280–298.

Mackay, H., & Gillespie, G. (1992). Extending the social shaping of technology approach: Ideology and appropriation. *Social Studies of Science, 22*(4), 68.

MacKenzie, D. (1996). Applying the sociology of knowledge to technology. In R. Fox (Ed.), *Technological change*. Amsterdam: Harwood.

Manning, P. (2001). Journalists and news production. In *News and news sources. A critical introduction* (pp. 50–80). London: Sage.

Manovich, L. (2001). *The language of new media*. Cambridge, MA: MIT Press.

Martin, S.E. (1998). How news gets form paper to its online counterpart. *Newspaper Research Journal, 19*(2), 64–73.

Marvin, C. (1988). *When old technologies were new: Thinking about electric communication in the late nineteenth century*. New York: Oxford University Press.

Massey, B., & Levy, M. (1999). Interactivity, on-line journalism, and English-language web newspapers in Asia. *Journalism and Mass Communication Quarterly, 76*(1), 138–151.

Matheson, D. (2004). Negotiating claims to journalism: Webloggers orientation to news genres. *Convergence, 10*(4), 33–54.

———. (2007) In search of popular journalism in New Zealand. *Journalism Studies, 8*(1), 28–41.

———. (in press). What the blogger knows. In M. Tremayne (Ed.), *Journalism and citizenship: New agendas*. London: Routledge.

McCombs, R. (2003). *Overcoming barriers to producing multimedia in online newsrooms.* Master thesis, University of Minnesota.

McLuhan, M. (1964). *Understanding media: The extensions of man.* New York: Signet.

McManus, J. (1994). *Market-driven journalism.* Thousand Oaks, CA: Sage.

McPhail, T. (2006). *Global communication: Theories, stakeholders, and trends.* Malden, MA: Blackwell.

McQuail, D. (1997). *Audience analysis.* London: Sage.

———. (2005). *McQuail's mass communication theory.* London: Sage (5th ed).

Mehlen, M. (1999). Die Online-Redaktionen deutscher Tageszeitungen. Ergebnisse einer Befragung von Projektleitern [Online newsrooms of German dailies. Results of a survey of project leaders]. In C. Neuberger & J. Tonnemacher (Eds.), *Online: Die Zukunft der Zeitung?* [Online: The future of the newspaper?] (pp. 88–123). Opladen, Germany: VS Verlag.

Meier, K. (2007). Innovations in Central European newsrooms: Overview and case study. *Journalism Practice, 1*(1), 4–19.

Meyer, K. (2005). *Crossmediale Kooperation von Print- und Online-Redaktionen bei Tageszeitungen in Deutschland* [Cross-media cooperation of print and online editorial staff of newspapers in Germany]. Munich, Germany: Herbert Utz.

Meyers, O. (2002). *A communication problem. Journalism studies, communication research and the production of news: The development of a rift.* Unpublished paper, University of Haifa, Israel. Retrieved from: http://smart.huji.ac.il/articles/Oren%20Meyers.pdf

Miller, G., Dinghall, R., & Murphy, E. (2004). Using qualitative data and analysis: Reflections on organizational research. In D. Silverman (Ed.), *Qualitative research: Theory, method and practice* (pp. 325–341). London: Sage (2nd ed.).

Monteiro, E. (2000). Actor-network theory and information infrastructure. In C. U. Ciborra et al., *From control to drift: The dynamics of corporate information infrastructures* (pp. 71–83). Oxford: Oxford University Press.

Moragas, M. de, Domingo, D., & López, B. (2002). Internet and local communications: First experiences in Catalonia. In N. Jankowski & O. Prehn (Eds.), *Community media in the information age.* Cresskill, NJ: Hampton Press.

Mortenson, T., & Walker, J. (2002). Blogging thoughts: Personal publication as an online research tool. In A. Morrison (Ed.), *Researching ICTs in context* (pp. 249–279). Oslo: InterMedia Report.

Mosco V. (2004). *The digital sublime: Myth, power and cyberspace.* Cambridge, MA: MIT Press.

Murley, C., & Roberts, C. (2005). Biting the hand that feeds: Blogs and second-level agenda setting. Paper presented at the *Media Convergence Conference,* Provo, UT.

Neuberger, C. (2000). Renaissance oder Niedergang des Journalismus? Ein Forschungsüberblick zum Online-Journalismus [Renaissance or decline of journalism? A research overview on online journalism]. In: K.-D. Altmeppen, H.-J. Bucher, & M. Löffelholz (Eds.), *Online-*

Journalismus. Perspektiven für Wissenschaft und Praxis [Online journalism. Perspectives for science and practice] (pp. 15–48). Opladen, Germany: VS Verlag.

———. (2002). Online-Journalismus: Akteure, redaktionelle Strukturen und Berufskontext [Online journalism: Actors, editorial structures, occupational context]. *Medien & Kommunikationswissenschaft, 50*(1), 102–114.

———. (2003). Onlinejournalismus: Veränderungen, Glaubwürdigkeit, Technisierung [Online journalism: Changes, credibility, technology-zation]. In *Media Perspektiven 3,* 131–138.

Neuberger C., Tonnemacher, J., Biebl, M., & Duck, A. (1998). Online, the future of newspapers? Germany's dailies on the World Wide Web. *Journal of Computer-Mediated Communication, 4*(1). Retreived from: http://jcmc.indiana.edu/vol4/issue1/neuberger.html

Neuberger, C., & Tonnemacher, J. (Eds.). (1999). *Online: Die Zukunft der Zeitung?* [Online: The future of the newspaper?]. Opladen, Germany: VS Verlag.

Oblak, T. (2005). The lack of interactivity and hypertextuality in online media. *Gazette, 67*(1), 87–106.

Ogbonna, E., Harris, L.C. (2006). Organizational culture in the age of the Internet: An exploratory study. *New Technology, Work and Employment, 21*(2), 162–175.

Oram, H. (1993). *Paper Tigers.* Belfast, U.K.: Appletree.

Pappacharissi, Z. (2006). Audiences as media producers: Content analysis of 260 blogs. In M. Tremayne (Ed.), *Blogging, citizenship and the future of media* (pp. 21–38). New York: Routledge.

Paterson, C. (2007). International news on the Internet: Why more is less. *Ethical space: The International Journal of Communication Ethics, 4*(1), 57–66.

Patterson, T.E., & Donsbach, W.G. (1996). News decisions: Journalists as partisan actor. *Political Communication, 13,* 455–468.

Paulussen, S. (2004). Online news production in Flanders: How Flemish online journalists perceive and explore the internet's potential. *Journal of Computer-Mediated Communication, 9*(4). Retrieved from: http://jcmc.indiana.edu/vol9/issue4/paulussen.html

Pavlik, J.V. (2000). The impact of technology on journalism. *Journalism Studies, 1*(2), 229–237.

———. (2001). *Journalism and new media.* New York: Columbia University Press.

Pedersen, T. (2006). *A study of the concept of interactivity as it applies to online newspapers.* Master thesis, West Virginia University.

Peterson, R.A., & Anand, N. (2004). The production of culture perspective. *Annual Review of Sociology, 30,* 311–334.

Pickard, V. W. (2006). Assessing the radical democracy of Indymedia: Discursive, technical, and institutional constructions. *Critical Studies in Media Communication, 23,* 19–38.

Pinch, T. (1996). The social construction of technology: A review. In R. Fox (Ed), *Technological change.* Amsterdam: Harwood.

————. (2001) Why do you go to a music store to buy a synthesizer: Path dependence and the social construction of technology. In R. Garud & P. Karnoe (Eds.), *Path dependence and creation.* Mahwah, NJ: Erlbaum.

Pottker, H. (2003). News and its communicative quality: The inverted pyramid: When and why did it appear? *Journalism Studies, 4*(4), 501–511.

Powers, S. (2003, February 23) Google is not God, Webloggers are not Capital-J journalists. *Burningbird.* Retrieved from: http://weblog.burningbird.net/archives/2003/02/23/

Project for Excellence in Journalism (2007). *The state of the news media 2007: An annual report on American journalism.* Washington, DC: Project for Excellence in Journalism. Retrieved from: http://www.stateofthemedia.org/2007

Pryor, L. (2002). The third wave of online journalism. *Online Journalism Review.* Retrieved from: http://www.ojr.org/ojr/future/1019174689.php

Puijk, R. (1990). *Virkeligheter i NRK.* [Realities in NRK]. Unpublished manuscript.

————. (2004). Ethnographic research in media organizations: Some methodological and ethical thoughts. Paper presented at the 2nd *Symposium New Research for New Media: Innovative research methodologies,* Tarragona, Catalonia. Retrieved from: http://www.makingonlinenews.net/docs/puijk_nrnm2004.pdf

————. (2007). Time and timing in multimedia production. In G. Lekakos, K. Chorianopoulos, & G. Doukidis (Eds.), *Interactive digital television: technologies and applications.* Hershey, PA: IGI.

Quandt, T. (2003). Towards a network approach of human action: Theoretical concepts and empirical observations in media organizations. Paper presented at the annual meeting of the *AEJMC*, Kansas City, MO.

———— (2005). *Journalisten im Netz* [Journalists in the Net]. Opladen, Germany: VS Verlag.

———— (2007). Network theory and human action. Theoretical concepts and empirical application. In A. Hepp, F. Krotz, S. Moores, & C. Winter (Eds.), *Network, connectivity and flow. Key concepts for Media and Cultural Studies.* New York: Hampton Press.

Quandt, T., Altmeppen, K.-D., Hanitzsch, T., & Loeffelholz, M. (2003). Online journalists in Germany 2002. Paper presented at the annual meeting of the *AEJMC*, Kansas City, MO.

Quandt, T., Löffelholz, M., Weaver, D., Hanitzsch, T., & Altmeppen, K.-D. (2006). American and German online journalists at the beginning of the 21st century. A bi-national survey. *Journalism Studies, 7*(2), 171–186.

Quandt, T., & Singer, J. (in press). Convergence and cross-platform content production. In K. Wahl-Jorgenson & T. Hanitzsch (Eds.), *Handbook of Journalism Studies.* Mahwah, NJ: Lawrence Erlbaum.

Quinn, S. (2005). *Convergence journalism.* New York: Peter Lang.

Quittner, J. (1995). The birth of way new journalism. *HotWired*. Retrieved from: http://web.archive.org/web/19990503195745/http://www.hotwired.com/i-agent/95/29/waynew/waynew.html

Reese, S. (1993). The media sociology of Herbert Gans: A Chicago functionalist. Paper presented at the annual conference of the *International Communications Association*, Sydney, Australia.

Rogers, E. (2003). *Diffusion of innovations (5th ed)*. New York: Free Press.

Rintala, N. (2005). *Technological change and job redesign. Implications of quality of working life*. Doctoral thesis, Helsinki University of Technology, Finland. Retrieved from: http://lib.tkk.fi/Diss/2005/isbn9512275139

Rosen, J. (2006). The people formerly known as the audience. *PressThink*. Retrieved from: http://journalism.nyu.edu/pubzone/weblogs/pressthink/2006/06/27/ppl_frmr.html

Rühl, M. (1998). Von fantastischen Medien und publizistischer Medialisierung [On fantastic media and the media-tization of publishing]. In B. Dernbach, M. Rühl, & A.M. Theis-Berglmair (Eds.), *Publizistik im vernetzten Zeitalter. Berufe, Formen, Strukturen* [Publishing in a network age. Jobs, forms, structures] (pp. 95–107). Opladen, Germany: VS Verlag.

Ryfe, D.M. (Ed.). (2006). New institutionalism and the news. *Political Communication, 23*(2).

Schaffer, J. (2007) *Citizen media: Fad or the future of news?* Maryland: J-Lab. Retrieved from: http://www.j-lab.org/citizen_media.pdf

Schiller, H. (1989). *Culture, inc.: The corporate takeover of public expression*. New York: Oxford University Press.

Schlesinger, P. (1980). Between sociology and journalism. In H. Christian (Ed.) *Sociology of the Press and Journalism*. Keele, UK: Univ. of Keele.

———. (1987). *Putting 'reality' together: BBC News*. London: Routledge (2nd ed).

Schmitt, C. (1998). Nachrichtenproduktion im Internet. Eine Untersuchung von MSNBC Interactive [News production on the Internet. A study on MSNBC Interactive]. In J. Wilke (Ed.), *Nachrichtenproduktion im Mediensystem* [News production in the media system] (pp. 293–330). Köln: Böhlau.

Schreiber, W. (1977). *The impact of automation technology on American daily newspapers and editorial division employees*. Doctoral dissertation, California State University.

Schudson, M. (2000). The sociology of news production revisited (again). In J. Curran & M. Gurevitch (Eds.), *Mass media and Society*. London: Edward Arnold (3rd ed.).

———. (2003). *Sociology of News*. New York: Norton.

Schultz, T. (2000). Mass media and the concept of interactivity: An exploratory study of online forums and reader e-mail. *Media, Culture & Society, 22*, 205–221.

Schütz, A. (1981). *Der sinnhafte Aufbau der sozialen Welt. Eine Einleitung in die verstehende Soziologie* [The phenomenology of the social world]. Frankfurt, Germany: Suhrkamp (2nd ed.).

————. (2002). *Theorie der Lebenswelt. Zur kommunikativen Ordnung der Lebenswelt* [The structures of the life-world]. Konstanz, Germany: UVK (vol. 2 of the revised ed.).

Schwartzman, H.B. (1993). *Ethnography in organizations*. London: Sage.

Scott, D.T. (2006). Pundits in muckrakers' clothing. In M. Tremayne (Ed.) *Blogging, citizenship and the future of media* (pp. 39–57). New York: Routledge.

Shoemaker, P.J. (1991). *Gatekeeping*. Newbury Park, CA: Sage.

Shoemaker, P.J., & Reese, S.D. (1996). *Mediating the message: Theories of influence on mass media content*. White Plains, NY: Longman Publishers (2nd ed.).

Silcock, B., & Keith, S. (2006). Translating the tower of babel? Issues of definition, language and culture in converged newsrooms. *Journalism Studies, 7*(4), 482–510.

Singer, J.B. (1997). Still guarding the gate? The newspaper journalist's role in an online world. *Convergence, 3*(1), 72–89.

————. (1998). Online journalists: Foundations for research into their changing roles. *Journal of Computer Mediated Communication, 4*(1). Retrieved from: http://jcmc.indiana.edu/vol4/issue1/singer.html

————. (2003). Who are these guys? The online challenge to the notion of journalistic professionalism. *Journalism, 4*(2), 139–163.

————. (2004a). More than ink-stained wretches: The resocialization of print journalists in converged newsrooms. *Journalism & Mass Communication Quarterly, 81*(4), 838–856.

————. (2004b). Strange bedfellows? Diffusion of convergence in four news organizations. *Journalism Studies, 5*(1), 3–18.

————. (2005). The political j-blogger: 'Normalizing' a new media form to fit old norms and practices. *Journalism, 6*(2), 173–198.

————. (2006a). Partnerships and public Service: Normative issues for journalists in converged newsrooms. *Journal of Mass Media Ethics, 21*(1), 30–53.

————. (2006b). The socially responsible existentialist. *Journalism Studies, 7*(1), 2–18.

Singer, J.B., Tharp, M., & Haruta, A. (1999). Online staffers: Superstars or second-class citizens? *Newspaper Research Journal, 20*(3), 29–47.

Soloski, J. (1997). News reporting and professionalism: Some constraints on thereporting of news. In D. Berkowitz (Ed.), *Social meanings of news* (pp. 138–154). Thousand Oaks, CA: Sage.

Spradley, J. P. (1980). *Participant observation*. New York: Holt, Rinehart & Winston.

Stake, R.E. (2005). Qualitative case studies. In N.K. Denzin & Y.S. Lincoln (Eds.), *The Sage handbook of qualitative research* (pp. 443–466). Thousand Oaks, CA: Sage Publications (3rd ed.).

Sullivan, C. (2004) Blogging's power to change journalism. *Editor & Publisher*.

Sun, X. (2001). *An orchestra of voices: Making the argument for greater press freedom in the People's Republic of China*. Westport, CT: Praeger.

Thalhimer, M. (1994). Hi-tech news or just shovelware? *Media Studies Journal, 8*(1), 41–51.

Thelen, G. (2002, July/August). Convergence is coming. *Quill,* p. 16.

Toong, J.W. (2005). Publish to perish: Regime choices and propaganda impact in the anti publications campaign. *Journal of Contemporary China, 14*(44), 507–523.

Trammell, K.D., & Keshelashvili, A. (2005). Examining the new influencers: A self-preservation study of A-list blogs. *Journalism and Mass Communication Quarterly, 82*(4), 968–982.

Tuchman, G. (1978). *Making news. A study in the social construction of reality.* London: The Free Press.

———. (2002). The production of news. In K.B. Jensen (Ed.), *Handbook of media and communication research: Qualitative and quantitative research methodologies.* Florence, KY: Routledge.

Von Hippel, E. (2005). *Democratizing innovation.* Cambridge, MA: MIT Press.

Wall, M. (2005). Blogs of war: Weblogs as news. *Journalism, 6*(2), 153–172.

———. (2006). Blogging Gulf War II. *Journalism Studies, 7*(1), 111–126.

Wang, H. (2006). To negotiate, not to irritate: A strategy of doing investigative reporting in China under political pressure. Paper presented at the *Workshop on Media Politics and Investigative Journalism in China,* Lund, Sweden.

Warner, M. (1970). Decision-making in network television news. In J. Tunstall (Ed.), *Media Sociology.* London: Constable.

Wellman, B. (2004). The three ages of internet studies: Ten, five and zero years ago. *New Media & Society, 6*(1), 123–129.

Wilke, J., & Joho, C. (2000). Journalistische Arbeitsweisen in Internetredaktionen am Beispiel des ZDF [Journalistic work in Internet newsrooms. The example of the ZDF]. In: K.-D. Altmeppen, H.-J. Bucher, & M. Löffelholz (Eds.), *Online-Journalismus: Perspektiven für Wissenschaft und Praxis* [Online journalism: Perspectives for science and practice] (pp. 95–106). Opladen, Germany: VS Verlag.

Williams, F., Rice, R.E., & Rogers, E.M. (1988). *Research methods and the new media.* New York: Free Press.

———. (1981). Communications technologies and social institutions. In R. Williams (Ed.), *Contact: Human communication and its history.* New York: Thames & Hudson.

Williams, R. (2003). *Television: Technology and cultural form.* London: Routledge (revised ed.).

Williams, R., & Edge, D. (1996). The social shaping of technology. *Research Policy, 25,* 865–899.

Willis, P., & Trondman, M. (2000). Manifesto for ethnography. *Ethnography, 1*(1), 5–16.

Wilson, T., Hamzah, A., & Khattab, U. (2003). The 'cultural technology of clicking' in the hypertext era: Electronic journalism reception in Malaysia. *New Media & Society, 5*(4), 523–545.

Wimmer, R., & Dominick, J. (2003). *Mass media research: An introduction.* Belmont, CA: Wadsworth (7th ed.).

Winston, B. (1996). *Misunderstanding media.* Cambridge, MA: Harvard University Press.

———. (1998). *Media technology and society. A history: from the telegraph to the internet.* London: Routledge.

WAN: World Association of Newspapers (2006). Outsourcing. *Future of the Newspaper Project Report, 5.*

Xu, W. (2007). *Chinese cyber nationalism: Evolution, characteristics, and implications.* Plymouth, UK: Rowman & Littlefield.

Ytreberg, E. (2002). Selvspill i radio: Mamarazzi's ukonvensjonelle populærjournalistikk. [Self play in radio; Mamarazzi's unconventional popular journalism]. Unpublished manuscript.

Yu, H. (2007). The graying of Chinese blogosphere: Bloggers strike a nerve. Paper presented at the *International Convention of Asia Scholars,* Kuala Lumpur, Malaysia.

Zhang, C., & Ni, J. (Eds.). (2000). *Guojia Xinxi Anquan Baogao* [Report on National Information Security]. Beijing: Renmin Chubanshe [People's Publishing House].

Zhong, Z. (Ed.). (2002). *Wangluo xinwen xue* [Internet journalism]. Beijing: Beijing Daxue Chubanshe [Beijing University Press].

Zhou, Y. (2006). *Historicizing online politics: Telegraphy, the Internet and Political Participation in China.* Stanford, CA: Stanford University Press.

Zingarelli, M. (2000). *Surfing the wave of flux: A journey into how conventional media are adapting to meet the demands of the new online medium.* Master's thesis, Carleton University, Canada.

Zollman, P.M. (2001). *Convergence or cooperation? Cooperation is best, for now.* Retrieved from International Newspaper Marketing Association: http://www.inma.org/members/perspective.cfm?col=161

Contributors

Jody Brannon (jody@jodybrannon.com) began working in online news in 1995 for the *Washington Post* followed by nearly five years at *USAToday.com* as executive producer. In 2006 she joined Microsoft's *MSN.com*. Her doctoral research examined online newsrooms and she has taught in the American University's graduate program in interactive journalism. She is a board member of the Online News Association, Knight-Batten Awards for Innovation in Journalism and the University of Washington's Graduate Program in Digital Media. Her website is www.jodybrannon.com.

Axel Bruns (a.bruns@qut.edu.au) is a Senior Lecturer at Queensland University of Technology (Australia), author of several books about the growing trend towards user-led content creation (*Gatewatching: Collaborative Online News Production*; *Blogs, Wikipedia, Second Life and Beyond: From Production to Produsage*) and coeditor of *Uses of Blogs*. His research blog is www.snurb.info.

Anthony Cawley (anthony.cawley2@mail.dcu.ie) is a research officer in the Center for Society, Technology and Media at Dublin City University (Ireland). His doctoral research studied innovation in Ireland's digital content industry. He has published in international journals and his research interests include content innovation, consumption of media technologies, and journalism.

Vinciane Colson (vcolson@ulb.ac.be) is a doctoral candidate in the Department of Information and Communication Sciences at the Université Libre de Bruxelles (Belgium). Her research interests are media convergence and science journalism.

Mark Deuze (mdeuze@indiana.edu) holds a joint appointment at Indiana University's Department of Telecommunications in Bloomington (USA) and as Professor of Journalism and New Media at Leiden University (The Netherlands). He has published five books—*Media Work* (Polity Press, 2007) is the most recent. He has also published articles in numerous journals, and runs a weekly radio show called Global Riffs (www.wiux.org). His weblog is deuze. blogspot.com.

David Domingo (david@dutopia.net) is Visiting Assistant Professor in the School of Journalism and Mass Communication at the University of Iowa (USA) and Assistant Professor in the Communication Department at Universitat Rovira i Virgili in Tarragona (Catalonia, Spain). His research focuses on the development of online journalists working routines and values, and their adoption of convergence and audience participation. He was president of the Catalan Online Journalism association (www.gpd.cat) from 2004 to 2006. His blog is www.dutopia.net.

Edgardo Pablo García (edgarcia307@hotmail.com) is a doctoral candidate at the University of Westminster (London) and Senior Lecturer at Universidad Argentina de la Empresa and Universidad de Belgrano in Buenos Aires. He has been a Lecturer at Aberdeen Business School and Visiting Lecturer at the University of Westminster. His main research interests are online journalism, political communication, and global media.

François Heinderyckx (Francois.Heinderyckx@ulb.ac.be) is a professor in the Department of Information and Communication Sciences at the Université Libre de Bruxelles (Belgium), where he is chair of the Masters in Information and Communication. He is president of the European Communication Research and Education Association (ECREA) and one of the project leaders of the Institut des Sciences de la Communication du CNRS (France). His research interests include journalism and new media, political communication, and ICTs.

Johan Lagerkvist (Johan.Lagerkvist@ui.se) is a research fellow and China specialist with the Swedish Institute of International Affairs in Stockholm. His research interests concern Chinese domestic and foreign policy, and the role of the mass media and the Internet in expanding the public sphere in East and Southeast Asia. He has published articles in *China Information, China: An International Journal,* and *Journal of Contemporary China.*

John Latta (jlatta@ua.edu) is a doctoral candidate in the College of Communication and Information Sciences at the University of Alabama (USA). He is a trade publication editor and formerly a journalist for dailies and weeklies in New Zealand, Australia, the USA, and the UK.

Wilson Lowrey (wlowrey@ua.edu) is an Associate Professor in the College of Communication and Information Sciences at the University of Alabama (USA). His research focuses on the sociology of news work and he has been published in a number of journals, including *Journalism & Mass Communication Quarterly*, *Journalism*, and *Journal of Media Economics*.

Chris Paterson (c.paterson@leeds.ac.uk) is a Senior Lecturer with the Institute for Communication Studies at the University of Leeds (UK). In 2004 he co-edited *International News in the 21st Century*. Paterson is an adviser to News-desk.org, and is co-founder of the Working Group on Media Production Analysis of the IAMCR.

Roel Puijk (roel.puijk@hil.no) is a social anthropologist and is currently professor in the Department of Television Production and Film Studies at Lillehammer University College (Norway). He has done research on television and cross-media production, the production and reception of the Olympic Games, and in other fields such as tourism and youth culture. He is currently the leader of the research project, "Television in a Digital Environment" (tide.hil.no).

Thorsten Quandt (thorstenquandt@t-online.de) is a Junior Professor with the Institut für Publizistik- und Kommunikationswissenschaft at the Free University Berlin (Germany). He has edited and co-published more than 50 articles and books, including *Journalism Research: An Introduction* (together with Thomas Hanitzsch, forthcoming). Currently, he is the chair of the research network, "Integrative Theories in Communication Studies," the chair of the Journalism Division in the German Communication Association (DGPuK), and an editor for the theory section of the international journal, *Journalism Studies*.

Jane B. Singer (JBSinger@uclan.ac.uk) is the Johnston Press Chair in Digital Journalism at the University of Central Lancashire (UK) and an Associate Professor in the School of Journalism and Mass Communication at the University of Iowa (USA). She is the coauthor of *Online Journalism Ethics: Traditions and Traditions* (M.E. Sharpe, 2007), and has been writing about journalists' transition to the digital media environment since the mid-1990s.

Index

3cat24.cat (news portal), 114–9, 122, 126

ABC News (broadcaster), 100–110
access (see ethnography)
action theory, 79
actor-network theory, 23–4, 28
anthropology of technology, 24–5
archive, news, ix, 29–31, 38–9, 58–9
ARD (broadcaster), 81
Argentina, 61–63, 66, 69
Associated Press (news agency), 103
audience (also see participatory journalism)
    feedback, 37–8, 119, 131, 196
    participation, 129, 170, 173–4, 201, 208
    conception of the, 26, 179, 180, 184, 185,
    188, 194–5, 209
Australia, 171–73
Australian, The (newspaper), 171–3, 176–9,
    184
autonomy of the online newsroom (see
    online–offline newsroom relationships)

BBC (broadcaster), 160–3
Belgium, 145
bloggers, 129, 134, 137–8, 172, 177, 180–2,
    185–97, 206
Boczkowski, Pablo, 2, 4, 15–20, 24–6, 114,
    119–22, 146, 159–61, 164
breaking news, 47, 50–2, 55–9, 66, 71, 99,
    107, 115, 117, 123

Catalonia, 113–4, 119
CCMA (broadcaster), 126
CCTV (broadcaster), 130–1, 135, 139
change, organizational, 19, 29, 33, 124 (also
    see technological change)
China, 127–41

citizen journalism, 1, 132, 95, 181, 201
Clarín (newspaper), 61–74
CNN (broadcaster), 8, 187
collaboration (see online-offline newsroom
    relationships)
communication history, 16, 20–2
competition, ix, 31, 39, 115, 125, 147, 168
    with other online media, 23, 26, 63,
    86, 90, 115, 136, 174, 208
    between print and online editions, 51, 63,
    73–4, 120, 132, 144–53
    within converged newsrooms, 158, 162,
    165 (also see online-offline newsroom
    relationships)
computer-assisted reporting (CAR), 200
constructivist research, 16–20, 25–7
content analysis, 2, 16, 164, 165, 182
content management systems (CMS), 84, 89,
    95, 122–4, 204–5
convergence, 83, 143–53, 157–70, 201, 206
    continuum, 157, 165
    dimensions, 151–3
Cottle, Simon, 2–3, 18, 29–30, 160–3
cross-platform production, 31, 157, 161,
    167 (also see convergence)
cultural resistance, 159

Daily Telegraph (newspaper), 145
deadlines, 46, 57, 71, 205
    lack of deadlines, x, 6, 86
desk journalism, 7, 67, 72–3, 86, 102, 144,
    204
Deuze, Mark, 2, 4, 7, 16–7, 115, 131, 144,
    150, 158, 200–13
Diari de Tarragona (newspaper), 114
diffusion of innovations, 100, 111, 163,
    167–9

digital myth, 3
digitization, 3, 32, 34, 36, 145, 201
disruptive technology, ix, 201
Dongfang (news portal), 130–139

El Periódico de Catalunya (newspaper), 114
email, use of, 35–8, 40, 57, 75, 148–9, 190
ethnography
    definition, 4–5, 158–62
    benefits, 5, 25, 161
    methods, 5, 29–44, 147, 157
    access, 8, 30, 32–37, 40, 45, 62
    interviewing, 5, 8 , 24, 78, 158, 162
    observation, 2, 5, 18, 158
    role of the researcher, 9, 62, 159
Europe, 15, 31, 144, 166

FAZ.net (news portal), 80–3, 89
Financial Times (newspaper), 145

Gans, Herbert, 3, 8, 45
gatekeeping, 38, 146, 172–8
gatewatching, 177–8
Germany, 77–9, 81, 96
Google, 138, 178, 182
Guardian (newspaper), 49, 144

hyperlocal, 181
hypertext, 4, 16, 119, 201

immediacy, 15, 47, 51–2, 56–7, 86, 106, 109,
    115–6, 119, 121, 205
impression management, 187
industrial journalism, 174, 179, 183
inhabited institutions, 203
innovation, 119, 123–4, 167, 209 (also see
    technological innovation and diffusion
    of innovations)
intellectual property, 147, 205
interactivity, 4, 16, 37, 63, 102, 109, 115,
    119, 201 (also see audience participation)
Internet as mass medium, 21, 54
Intranet, 32, 35, 204

Ireland, 45, 49, 51
Ireland.com (news portal), 45, 50–4, 59
Irish Times (newspaper), 45–59

Japan, 128, 130, 135–6, 139

Kuro5hin (weblog), 181

La Libre Belgique (newspaper), 145–54
laMalla.net, 114–6, 119–20, 125
law of radical potential suppression, 22, 125
libel, 47, 205
liquid news, 205–6

mainstream journalism, 179, 186–7
managerial decisions, 23, 27, 33, 51, 53, 64,
    104, 106–9, 120–1, 125, 144, 149–53,
    161
media logic, 26, 110, 193–6
media routines (see routines)
methodology (see ethnography, methods)
Microsoft, 138
mobile phone, 31, 153
multimedia, 16, 110, 201
    production, 29, 73, 89, 102, 153, 162
    news, 58, 65, 67, 73, 119
multi-skilling, 108, 161–3, 207 (also see
    convergence)

nationalism, 128, 135–9
networked public sphere, 183
Netzeitung (news portal), 80–3, 89. 96
 new media, definition, 28
New York Times (newspaper), 49, 146, 170,
    187
news agencies, 3, 7, 37, 55, 62–3, 68–72,
    82–4, 86, 89–90, 93, 95, 97, 99,103,
    116–8, 127, 139, 150, 152, 161, 176–7
newsgathering online, 37, 47, 55, 118, 161,
    167, 201
newshole, 115, 175
newsroom, online
    culture, 100, 157, 160, 163, 200–3, 205

organizational structure, ix, 18, 113, 146, 167, 202, 206

physical structure, 47, 50, 53, 81–5, 102, 148

normative research, 15–6

Norway, 39

NPR (broadcaster), 100–7, 110

NRK (broadcaster), 29–36, 39

objectivity, 131, 162, 188 (also see online journalists, values)

OhmyNews (news portal), 181

online-offline newsroom relationship
control, 64–8, 70, 73, 81, 121, 149
collaboration, 56, 73, 116, 120, 146–7, 149, 153, 207
editorial meetings (see routines, meetings)
editorial policies, 64–7, 69–70, 74, 144, 205

online journalists
demographic profile, 47, 53, 74, 80, 102, 146, 207
self-perception, 51, 53–54, 71–2, 74, 78, 89, 102, 131–132, 144, 150, 169
skills, 21, 26, 48–49, 58, 71, 102, 106–7, 122–3, 144, 160–1, 208
values, x, 3, 18, 26, 58, 105, 108, 113–4, 123–4, 131, 161–2, 176, 186, 188, 207–8

participatory journalism, 6, 16, 171, 174, 177–183 (also see citizen journalism)

passive journalism (see desk journalism)

PDA, 31

podcast, 31, 95, 151

print bias, 199

print newsrooms, 7, 39, 47, 50–4, 56–8, 61, 67–75, 84, 116, 144–53, 162, 168, 176, 199, 207

produsage, 189

professional identity, 16–17, 36, 51, 63, 72, 123, 131, 144, 148, 165, 204–6, 208 (also see online journalists)

propaganda, 129–30, 132, 136–40

public relations, 3

public service, 31, 159, 161, 169

pundits, 172–3, 177, 180, 199

Qianlong (news portal), 130–9

quality of online news, ix, 6, 58, 101, 107–9, 122, 125, 148–9, 165–6, 168, 176–7, 192

regulation of the Internet, 129, 132–3, 140, 205

repurposing, 50–2, 56, 67, 84–6, 89, 100, 119, 205 (also see shovelware)

research, non-ethnographic, 16, 164–5 (also see ethnography)

resocialization, 146, 169

Reuters (news agency), 117

rhetorical closure, 22, 27, 74

Rosen, Jay, 170, 173, 185

routines
writing, 30, 46–7, 51–52, 86, 89–90, 105–8, 118, 133, 149, 153, 158, 192–193, 196
newsgathering, 7, 118, 167
editing, 35–6, 47–8, 51, 56, 67, 82–4, 89–90, 116–118, 131–132, 162, 205
editorial meetings, 30, 33–5, 40, 52–7, 67–8, 75, 148–9

RSS, 191, 197

Schlesinger, Philip, 2–3, 5, 8

Schudson, Michael, 3, 18, 26, 114

second-class journalism, 6, 53–4, 71–3, 116, 144–5, 162, 206–8

shovelware, 7–8, 66, 81, 83–4, 89, 97, 100

Sina (news portal), 130–140

Singer, Jane, 3–4, 16–7, 116, 134, 146–50, 159–62, 167–70, 186–7

Slashdot (weblog), 178, 181

social construction of technology studies, 22–4, 28

sociology
of knowledge, 19

of news production, 2–3, 18–9, 25, 30, 96, 100, 114, 168, 200, 203, 208
sources, 3, 37, 46–7, 50–1, 70–3, 90, 93, 95, 116–9, 127, 140, 176–9, 188, 191
Spain, 113–4, 117
special coverage of events, 66–7, 73, 116, 119–20, 136, 150
speed of work, 86, 95–6, 102, 109, 166, 207
Spiegel Online (news portal), 80–3, 96
streaming, 31, 57–8, 119
supervening social necessity, 21
surveys, x, 2, 4, 16, , 157–8, 160, 165–6, 200
SVZonline (news portal), 80–4

tagesschau.de (news portal), 80–3, 89
Taiwan, 135
technological change, 16, 19–21, 31, 95, 159
technological choice, 24, 26, 27, 115
technological determinism, 4, 15–7, 19, 77, 100, 111, 161
technological innovation, 3, 16–26, 125
telephone, use of, 37, 46–7, 50–1, 64, 90, 95, 102, 116, 118, 148–9
text messaging (SMS), 36
theoretical frameworks, 20–5
training, 106, 109, 122, 151, 153, 167–8
transcription, 8
translation, 24, 26
triangulation, 165–6, 168

Tuchman, Gaye, 2–3, 30, 37, 188, 191, 202

United Kingdom, 10, 144
United States, x, 3–10, 15, 49, 99–101, 116, 120, 128, 146, 158, 165, 173, 181–2, 189
USA Today (news paper), 100–16, 119–20
user-generated content, 38 (also see participatory journalism)
utopian predictions, 3, 4, 115, 119–21, 125, 199, 205

video, 31, 34, 51, 57–9, 67, 71, 75, 101, 104, 119, 123, 158, 174
virtual ethnography, 181
Volkskrant (newspaper), 145

Washington Post (newspaper), 187
Web 2.0, 201
webcast, 57–8
weblogs, 95, 171, 186–7 (also see bloggers)
Wikipedia, 179
Winston, Brian, 21, 23, 125
wire service (see news agency)
word processor, 69, 84, 93, 95
work shifts, 66, 68, 71, 82, 96–7, 116

Xinhua (news agency), 139

Yahoo, 138, 141